THE
INTELLIGENT
COSMOS

Also, by Anthony J. Marolda

The Inventor and the Inventor's Son: The Two Isaac Adams
A biography

Howard to the Rescue!
A Children's Book

And four historical novels

The Raid on H.M.S. Vindicator

Purgatory Island

Operation Blinding Light

The Adventures of Luke Watson, M.D.

All books are available on Amazon and other book sellers.

Anthony J Marolda

February 2022

Graeme and Karin,

Here is how I kept busy during the Covid lockdown. It is based on a series of articles I wrote for the Gloucester Daily Times.

Best,

Tony

ANTHONY J. MAROLDA

February 2022

The Intelligent Cosmos

Intelligence: *Human, Animal, Artificial, and Extraterrestrial*

To Anne and Karin,

Anthony J. Marolda

with warm best wishes,

Tony

5

ISBN 978-0-9793099-4-6 hardcover

Book design by Anthony J. Marolda

Printed and bound in the United States of America
First Edition

Procyon Press
Gloucester, Massachusetts

For Matt and Ria,
the son and daughter
every father hopes to have.

With Matt and Ria when they were both students at Bowdoin College

Contents

THE INTELLIGENT COSMOS

Preface
A Quest for Understanding

This is a book born of the Covid Pandemic. For several years, I had been authoring articles for the **Gloucester Daily Times** newspaper. The articles were based on assorted topics in science. I chose the subjects randomly, I thought.

When the pandemic hit, the Gloucester Times shrunk its staff as well as the size and number of issues of the paper. It appeared that they would stop publishing my articles, so I began to look for another outlet for my science writings.

Dr. Andrew Wittkower lives in Rockport, Massachusetts, not far from my home in Gloucester. He is a former colleague from *High Voltage Engineering Corporation*. Coincidently, High Voltage had given the both of us our first positions as physicists.

Andrew had read most of my twenty-five, Gloucester Times articles and later suggested that I should collect them into a book. That sounded like a promising idea to fill my need for a writing outlet, so I made a list of the articles.

As I analyzed the topics, I noticed an overall theme which I did not realize was there. In one way or another, *many had to do with the idea of "intelligence"*. They included topics on human, animal, artificial and extraterrestrial intelligence. Suddenly, I saw that focus as the theme for a book. ***The Intelligent Cosmos*** was born.

The Contents

The **Table of Contents** for ***The Intelligent Cosmos*** is finely structured. Scanning it will provide a quick grasp of the range of topics covered in the book.

For the last one hundred years, we humans have searched deeply for an understanding of our own intelligence, what it is, how it evolved and how to measure it. And we have made great progress. Part I of this book, ***Human Intelligence***, presents the history of that search, and its results.

As you will note when you read this section, the quest to understand and measure human intelligence *had a dark side*. And this led to an overreaction by some people. The unfortunate effect was the slowing down of our progress in understanding *the role of nature vs. nurture in intelligence.*

However, the DNA and genetics revolutions occurred during the decade of the nineties, and the roles of nature vs. nurture in intelligence became clearer from a scientific perspective. *The result was an adjustment of some of those negative attitudes.*

Also covered in detail in this section is "consciousness", *one of the great mysteries of neuroscience.* Consciousness is fundamental to intelligence, and like intelligence, consciousness correlates with the functions of the brain. Several theories are presented about how this occurs, but there is no scientific consensus.

Charles Darwin, back in 1859, authored his famous book, **On the Origin of Species**. Long before we understood genetics, he proposed a theory of evolution, the basics of which hold up today. *Scientists became extremely interested in the role of evolution in the development of intelligence.* We explore this topic in Part II, *Animal Intelligence.*

The evolution of cognition is the idea that life on Earth has gone from organisms with little or no cognitive function to the wide range of cognitive abilities that we see in today's diverse creatures, including humans.

Comparative animal cognition provides a fascinating look into the varying levels of intelligence, and how it is expressed, in a wide range of creatures from the Great Apes to Cetaceans to Avians to Cephalopods.

I first became aware of *Artificial Intelligence* (**AI**) during the 1980s. At the time, I was working at the international consulting firm of Arthur D. Little, Inc. (ADL) helping companies develop their corporate strategies. ADL was located in Cambridge, Massachusetts and had on staff more than two thousand scientists and engineers from a wide range of disciplines. Some were specialists in the AI field.

During our work with clients who had an interest in AI, we would present the state-of-the-art for them and then help them to weave the technology into their corporate strategies.

As you will note in Part III, *Artificial Intelligence*, the development of AI had many stumbling blocks. Now, however, we are *on the verge of an intelligence explosion that will dramatically change the lives of everyone on planet Earth.*

As discussed in Part IV, the *Search for Extraterrestrial Intelligence* (SETI) is a topic of serious scientific interest. For example, Harvard University had a major SETI facility in the town of Harvard, Massachusetts, where I lived for many years. I became interested in the topic when I had an opportunity to take an astronomy course at the Harvard observatory and received a briefing by the astronomers on their SETI program.

UFOs and Flying Saucers have become associated with extraterrestrial intelligence in our popular culture. I grew up in the 1950s when the phenomena became widespread, and there was a great deal of talk about these objects. Soon, their discussion became a joke, and people who claimed to see such craft were ridiculed.

Now we know that the development of this attitude may have been the result of a plan by the Federal Government to hide the truth about these strange objects.

This situation lasted for over fifty years but has changed recently. In June 2021, the Pentagon produced a congressionally mandated report saying that, *after all, UFOs are indeed real and very mysterious*.

Most likely, the objects *do not originate in our civilization*. This does not mean, however, that they are, necessarily, of extraterrestrial origin. There are other, mysterious, Earth-based possibilities that need to be considered.

Finally, I conclude the book with Part V, **The Future of Humankind**. Up until now, the growth of intelligence on earth has been the result of the ultra-slow process of Darwinian evolution. *But the genetics revolution and the advent of genetic engineering will dramatically increase the speed of human change.*

Man-made modifications to the gene pool using these technologies will rid us of genetic diseases and even improve our intelligence and physical traits. And all within several generations, rather than thousands of years.

Perhaps a more important development that will further influence our evolution is the advent of an Artificial **Superintelligence** *(ASI).*

We are fast approaching the time when such a machine-based entity will be born. And *intelligence on Earth will explode to the point* **where current human intelligence will only represent a small fraction of the total.**

NASA and the private company, SpaceX, are planning on a multi-year program, starting in 2024, to build a permanent, self-sustaining colony on Mars. The colony could well be thriving by *the middle of the twenty-first century* with over 100,000 people living there.

And so this is the future of humankind, super intelligent humans *on their way to the stars and ultimately the cosmos. Over the following hundreds and thousands, and millions of years, we will be spreading the phenomena of intelligence and consciousness throughout the universe. And it is our destiny to do so.*

The Purposes of *The Intelligent Cosmos*

I had several intentions in authoring this book. *The first was to make it interesting and understandable to the average reader.* And yet to have it be scientifically accurate. *These are not always mutually achievable objectives.*

Based on the research I did for the background of the book, I cited in the **Endnotes** more than three hundred and fifty references to scholarly journal reports, news articles, and books. *These* *references are available to the more advanced reader who might be interested in digging deeper into a particular subject.*

Whenever I used technical terms in the text that I believed the average person would not necessarily understand, I *added a footnote* on the same page to explain the term. These are noted with roman numeral superscripts, as opposed to the Arabic numerals used to denote an Endnote.

As I was writing the text, I highlighted important phrases in italics. *My hope was that this technique would aid the reader to focus their attention on a key point.* It could also speed up the reading process.

I am a "visual" person and like the idea of illustrating a book to bring it to life. So, throughout the text, I have included relevant images. Most are portraits of the people, scientists, academics, and engineers, involved in the particular subject matter. Others are explanatory charts and graphs.

Yet another intention for this book is to make the reading of the material less daunting *by composing the topics so they can be read separately, and even randomly, depending on the time and interests of the reader.*

There are some sections of the book, for example the discussion of "Quantum Mechanics and Consciousness", which some readers may find too challenging to absorb or are of low interest. My suggestion is to not get bogged down. Just skip over these parts.

Let us say you opened the *Table of Contents* and saw the section on the *DNA Revolution*. And you noticed that, within that subject, there was a section on *Behavioral Genetics* that could be of special interest to you. That topic happens to be contained in four pages of text and can be understood without having to read everything that came before it.

Similarly, if you are more interested in **Extraterrestrial Intelligence**, Part IV of the book, than, say, **Animal Intelligence**, Part II, you might read the section on **Extraterrestrial Intelligence** first.

Part V is called the **Future of Humankind.** This Part draws on all of the science and technology developments discussed throughout the book *to examine the likely future of humankind.*

It is actually a good place to start the reading of the book. It deals with the continued evolution of humans, not by the slow process of Darwinian change, but by genetic engineering and our merger with technology.

And then points to the *destiny of humans*, to spread our consciousness and intelligence throughout the solar system, galaxy and then the rest of the Universe.

Acknowledgements

Back in the 1970s, when I first began writing on professional topics, I had to spend hours in libraries, like the one at the *Massachusetts Institute of Technology*, searching for books and scholarly articles in peer-reviewed journals that dealt with my topic.

There are over three hundred and fifty citations in the *End Notes* of **The Intelligent Cosmos**. If I had to search for all of those tomes and papers and articles the old-fashioned way, the book would have taken many more years to complete, *and probably would not have even been started.*

However, today, most scholarly papers are available instantaneously from the Internet. Relevant professional books are inexpensive and quick to obtain online through Amazon or Barnes and Noble.

The result is that the writing process is much easier, more efficient, and faster. I can spend more time digesting the material and considering its implications for my arguments.

Most importantly, I owe a large debt of gratitude to my wife, Maria. She supplied advice and encouragement throughout the work. And, as she has done on all of my writing efforts, she served as the book's primary editor. It is safe to say that, without her, **The Intelligent Cosmos** would not have been written.

About the Author

Science has been a life-long interest of mine. My education includes a master's degree in physics. So, my first job was as a physicist. I was fortunate that, when I graduated from university, it was during the time of the Cold War. As a result, the Federal Government was investing a great deal into Research and

Development. Consequently, some extremely interesting positions for a young physicist were available in the Boston area where I lived.

My first situation was with *High Voltage Engineering Corporation* of Burlington, Massachusetts, a *manufacturer of particle accelerators for nuclear physics research*. The accelerators were used in universities and government laboratories around the world.

High Voltage was formed by professors from M.I.T., including Dr. Robert Van de Graaf, the inventor of the Van de Graaf particle accelerator. Another founder was John Trump, the uncle of President Donald Trump. Dr. Trump's interest was in using electron beam accelerators for medical applications. The radiation therapy facility at Lahey Hospital and Medical Center, in the Boston area, is named for him.

My assignment while at High Voltage was to improve the performance of a Duoplasmatron ion source used in the particle accelerators. I had to become familiar with plasma physics and that became a life-long interest.

My second job was with Edgerton, Germeshausen & Grier (E.G.&G.) Inc. This was another Boston area firm that was formed by M.I.T. professors, including "Doc" Edgerton, the inventor of strobe lights and high-speed photography. His M.I.T. team included Kenneth Germeshausen and Herbert Grier. They had worked together on the *Manhattan Project during World War II* developing the timing and firing technology for the nuclear weapons, and then analyzing the weapons' effectiveness during the tests.

After the war, the government wanted the M.I.T. team to continue their efforts. So they formed E.G.&G., Inc. The company became heavily involved in all aspects of the country's nuclear weapons program. In addition to taking part in the design, testing and analysis of nuclear weapons effects, their business included the administrative management and operation of the Nuclear Test Site in Nevada, and, as it turned out, the management and operation of the adjacent, **Area 51**.

My assignment at E.G.&G. was with the Simulation Laboratory. My responsibility was the development of a megavolt, pulse power, electron beam accelerator to simulate the effects of a nuclear weapon on various materials.

Later, after I received an MBA from the Harvard Business School, I became a Vice President with the international consulting firm of Arthur D. Little, Inc. (ADL), in charge of their Management Counseling Section.

So, for more than thirty-five years of my consulting career, at ADL and my own firm, The Winbridge Group, I had an opportunity to work closely with major domestic and international firms involved in such fields as health care technology, defense systems, pharmaceuticals, industrial automation, semiconductors, electronics devices and components, and automotive technologies.

The subjects of our engagements were to collaborate with our clients' top management groups on the development of business unit and corporate strategies. Necessarily, we had to get deeply involved in understanding the clients' technology bases in order to incorporate that understanding into their plans.

Our ADL and Winbridge case teams included members who were expert in the technologies of the client. This was the strength of ADL since we had a staff of about 2,500 professionals where more than three quarters were experts in almost every technology imaginable.

The ADL technical staff, under contract to NASA, even developed some of the instruments used for experiments on the Apollo 11 moon landing. And another group originated and developed the concept of gigantic satellites in geostationary orbit to gather solar energy and beam it to Earth with microwaves.

The Part of this book on **Extraterrestrial Intelligence** includes a discussion of alien *Dyson Spheres*, which are related to the idea of giant, solar power satellites.

So, over my career, I was exposed to a wide range of sciences and technologies, mentored by experts, and had opportunities to take deep dives into them. Many of these areas of science and technology are related to the diverse topics of **The Intelligent Cosmos**. It made the writing easier. The end result, hopefully, is that it is also more interesting and informative for the reader.

Anthony J. Marolda

ANTHONY J. MAROLDA

Part I

Human Intelligence

I.   Human Intelligence

The Many Facets of Intelligence

Introduction

Without consciously thinking about it, we judge the people we meet on several factors. The first thing we notice is their physical appearance. In a matter of seconds, we know several things about them; male or female, relative height compared to ours, physical build, hair style and color, wearing glasses or not, the degree to which their face has pleasing features, the sound of their voice, the quality and style of their clothing, etc.

Later, as we begin to interact with them, we may intuit how "smart" they are. We listen to what they say, how they say it, and the value to us of their insights into everyday matters. And we make our instinctual estimate of their level of intelligence. Again, without necessarily even consciously thinking about it.

We may put our estimation of their intelligence in relation to ourselves. "Is the person smarter than me, about the same or on a lower level?" And this assessment has a lot to do with how we feel about them.

So, we all have a general understanding about what it means to be smart. We know it when we see it. And we have a good awareness of how smart we are compared to the people around us. *But understanding a person's "objective level of intelligence" is much more complex than that.*

The Merriam-Webster dictionary defines "intelligence" as:

1. the ability to learn or understand or to deal with new or trying situations,

2. the skilled use of reason,

3. the ability to apply knowledge to manipulate one's environment or to think abstractly as measured by objective criteria (such as tests),

4. mental acuteness or shrewdness.

This kind of definition is adequate for routine situations, but for a richer understanding, we need to dig deeper.

Fifty-two researchers into human intelligence collectively agreed on a definition of intelligence.[1]

> *"Intelligence is a very general, mental capability that, among other things, involves the ability to reason, plan, solve problems, think abstractly, comprehend complex ideas, learn quickly and learn from experience.*
>
> *"It is not merely book learning, a narrow academic skill, or test-taking smarts. Rather, it reflects a broader and deeper capability for comprehending our surroundings – 'catching on,' 'making sense' of things, or 'figuring out' what to do."*

Scholars who have studied intelligence, *sometimes called general cognitive ability*, have established that there are many components or facets of human intelligence. *First is the capacity to understand complex situations and identify the problems that confront us. Then to think abstractly about a difficulty, explore alternative solutions, and come up with a plan to implement an answer.*

To develop and carry out a successful plan requires the individual to have a particular set of skills. These include *the ability to learn from their experiences, apply logic and reasoning using this knowledge, be creative in developing practical possibilities, and then apply critical thinking and problem-solving skills to reach the final, successful solution.*

But how can a capability as complex as human intelligence be measured on an objective basis? Scientists have been working on this need for over a century.

How Intelligence is Measured

It all began in 1905. The government of France passed a law that all children must attend school. Since the officials were concerned with the need to identify children who might require extra assistance in learning, they commissioned psychologists Alfred

Binet and Theodore Simon to develop a test to accomplish this goal.

The result was the *Binet-Simon Scale*. It became the first intelligence test and was used for many years in France to evaluate the placement of children in the public schools.

The original form was expanded and revised. New versions were published in 1908 and 1911. These new forms were the result of extensive research and testing involving "normal" as well as mentally retarded examinees.

Then in 1916, Lewis Terman of Stanford University in California revised the Binet-Simon Scale for broader use in the United States.[2] Terman was a professor of educational psychology whose focus was on the teaching of the gifted child.

Terman authored a manual that presented translations and adaptations of the French items, as well as new ones that he had developed and tested between 1904 and 1915.

Terman's test is still in use today. The updated, fifth version is known as the Stanford-Binet 5 (SB5) and is one of the most widely used instruments to measure cognition.

The most common exam today for adults, however, is the Wechsler Adult Intelligence Scale (WAIS). In addition to the Stanford-Binet, there are others that may be employed by psychologists.[3] These include The Wechsler Intelligence Scale for Children (WISC), and The Woodcock-Johnson Tests of Cognitive Abilities.

Each test is somewhat different in terms of exactly what is being measured, how it is scored, and how the scores are interpreted. *But it has been found over the years that the results of the various exams are well correlated with each other.*

The term "IQ" was created in 1912 by German psychologist, William Stern of the University of Breslau. He wrote a book about his scoring method for intelligence tests and *called it "intelligenzquotient" or IQ in abbreviated form.*

All of the intelligence tests give a mental age in years and months, for the patient. This result is divided by the person's chronological age. **The subsequent fraction is multiplied by 100 to get the IQ score.**

In any of these IQ tests, the subject undertakes several different, cognitive tasks. These may include sections on vocabulary, verbal comprehension, spatial skills, abstract reasoning, concentration, memory, attention, and processing speed. *These capabilities represent the multiple components of human intelligence.*

In today's tests, the median raw score of the *norming population*[I] is defined as 100. And the individual's IQ score is found on the normal curve of the results to determine how they compare to the general population (*see the discussion on the Bell Curve in the next section*).

Wechsler Adult Intelligence Scale (WAIS) is a well-accepted instrument that provides such a single number for a person. But how can we depend on a few digits to define an individual's level of intelligence? It is because of the "g" factor.

An interesting finding by Psychologist Charles Spearman in 1927 was that *people who do well on one of the cognitive exams that make up an IQ test, say vocabulary, tend to do well on all the other components.* He postulated, based on this observation, that *there is an underlying,* **general intelligence factor** *that he termed,* "*g*".[4]

Charles Spearman

[I] **Norming** refers to the process of constructing norms or the typical performance of a group of individuals on a psychological or achievement assessment. Tests that compare an individual's score against the scores of groups are termed *norm-referenced assessments.*

Technically, it is *a variable that summarizes the correlations of performance between the different, specific cognitive tasks that, together, make up an IQ exam,* as shown in the diagram below.

The standard approach to calculating the "g" factor is to use a statistical technique called *exploratory factor analysis.*[5] *But the full-scale IQ score from the test battery has been found to be highly correlated with g factor scores,* **so the IQ score can be regarded as an estimate of g for that individual.**

In 1963, University of Illinois psychologist **Raymond Cattell found that "g" is composed of two components, g_f, or fluid intelligence, and g_c, or crystallized intelligence.**[6] He defined *fluid intelligence as the ability to solve novel, reasoning problems.* It depends on *innate reasoning ability,* and only minimally on prior education.

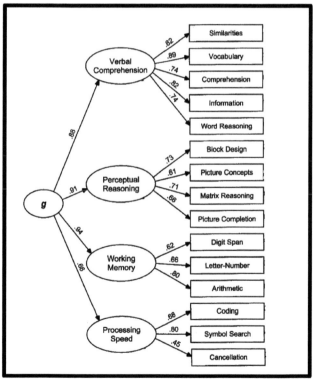

Example of outcome of hierarchical factor analysis of a WISC-IV IQ Test

Tasks that *measure fluid intelligence* include figure classifications, figural analyses, number and letter series, matrices, and paired associations. Most of the major intelligence tests, such as WISC and Woodcock-Johnson, include components that measure *fluid intelligence.*

Crystallized intelligence, on the other hand, *depends on learned procedures and knowledge.* Examples of tasks that measure crystallized intelligence are *vocabulary, general information, abstract word analogies, and mechanics of language.*

The Bell Curve

If you consider the IQ scores of a large sample of the population, the results tend to fall under a "Normal Curve".[7] This is the symmetrical, *bell-shaped curve that commonly describes the natural distribution of many, different human variables, including intelligence.* Other examples of human traits that fit under this type of curve include height, weight, and blood pressure.

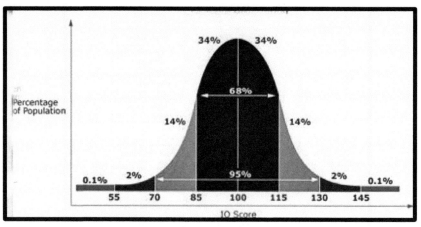

The Bell Curve - Normal Distribution of IQ Scores

In the case of the IQ test, the x-axis is the IQ score. The y-axis is the percentage of the population receiving that score. As shown in the diagram, the highest point on the curve, represents the average or the mean IQ score of the population, **which is normalized to be 100.** Then, we note from the *statistical characteristics of the Normal Curve* that 68% of the tested individuals have scores between 85 and 115. 90% are between 70 and 130.

Only a small fraction of people (about 2.5%) have a very low IQ below 70; or a very high IQ above 130. *People who have extremely high IQ's, of 160 or higher, are considered to be "geniuses".* For example, Albert Einstein, who is well-known for developing the Theory of Relativity, had an estimated IQ of about 160. That's about the same as the famous Astrophysicist, Stephen Hawking. Both scientists were in the top 0.1% of the population.

Long-term Life Outcomes

Many studies have been done that show that a person's IQ, as measured on the WAIS, or similar exams, has a lot to do with their success in life. The higher the IQ and "g" of an individual, the better they tend to perform in their educational experiences, their jobs after they finish school and even their health outcomes.

For example, scores on achievement exams in various subjects, and the grades received in school are both highly correlated with IQ scores. But as you might guess, the results that a student receives on achievement exams have a higher correlation with her IQ scores than the grades she receives in her coursework.

The reason may be that the achievement tests are more objective in measuring her knowledge and cognitive capabilities. Whereas grades also have some dependence on the teacher's subjective opinions of non-skill factors such as flexibility and her ability to apply those skills in a variety of different settings.

European longitudinal studies[II], one in Poland [8] and another in Sweden[9], examined the long-term success of youngsters based on IQ. The IQs of the students were measured when they were young, about ten or eleven years old, and again later in middle age. The objective was to see if there was a correlation of IQ score and objective indicators of success in life. The general finding in both European studies was that IQ was a relatively good predictor

---

[II] A longitudinal study is a research design that involves repeated observations of the same variables (e.g., people) over usually long periods of time.

of life success. The Swedish study, for example, found that there was a graded increase in income by IQ groups.

IQ and g are also positively correlated with other personal outcomes, including social and health. In 1979, the U.S. Bureau of Labor Statistics conducted a national, longitudinal survey of youth, age 14 to 22. The representative sample consisted of 12,686 young men and women. One of the original items tested in the sample was IQ. Each year thereafter, additional information was gathered on the sample. In 2009, twenty years later, a module, "Health at 40" was administered to the population. The unit included assessments of general health and depression, nine medically diagnosed conditions, and thirty-three common health problems.

A group of researchers from three UK universities evaluated a subsample of the group that had both their IQ measured in 1979 and had completed the "Health at 40" interview module in 2009. The data included 7,476 participants.[10]

The researchers found that there was an overall, positive correlation of higher IQ score with positive outcomes on most of the health measures included in the module. There were lower depression scores, significantly lower odds of having five of the nine diagnosed conditions, and fifteen of the thirty-three health problems.

However, from our personal experiences, we know that intelligence is a necessary, but not sufficient, trait for long-term success in life. So, what other behaviors or attributes do "successful" people need to have to achieve positive outcomes?

One key success factor identified by researchers is "grit".[11] *Grit is a characteristic often described as having perseverance and passion toward achieving important, long-term, personal goals.*

Professor Angela Duckworth is a psychologist at the University of Pennsylvania who found in her work that it takes grit as well as intelligence to realize positive life outcomes.

Duckworth did a ten-year, longitudinal study of 11,258 cadets at the U.S. Military Academy, West Point. Her findings suggest that success entails not only having talent but *also the sustained and focused application of that talent over time.*

Other important success factors include *learning from failure*[12] and *keeping an optimistic attitude* as you drive toward achieving your objectives[13].

It was found by Professor Robert C. Wilson, et al, of the University of Arizona, that having *some failures*, say 15% of the time, *was important to learning.*

If you cannot count on success every time you try something, you become uncomfortable. You find that you need to push yourself harder to achieve your goals. The researchers found that this need try harder is beneficial to learning and is, indeed, a key success factor.

Professor Uma Mishra, et al, of the Central University of Rajasthan, India, studied the *role of "optimism"* in job performance and satisfaction. The study was conducted on a sample of 346 employees from three large, public-sector banks situated in the Eastern part of India.

A person who is an optimist has positive expectations and faces life with a confident view. The results of the study showed that optimism was positively correlated with employee performance and job satisfaction.

Finally, Dr. Daniel Goleman, an author and science journalist, has proposed that *Emotional Intelligence leading to emotional competence,* is as critical for success in today's organizations as having a high IQ. *Emotional intelligence, or being intelligent about emotions,* is the capacity for recognizing one's own feelings and those of others, for motivating ourselves, and for *managing emotions* effectively in ourselves and others.[14]

Emotional competence is a learned ability, based on emotional intelligence, that contributes to effective performance at work. And it plays a leading role in the success of those employees who wish to reach leadership positions. Leadership is a dynamic, interpersonal process that incorporates a wide array of *cognitive as well as emotional competencies.*

IQ Measurement Controversy

The Rise and Fall of the Eugenics Movement

An original investigator of *the concept of the heritability[III]* of intelligence was Sir Francis Galton (1822-1911).   Galton was an English scientist and polymath, a man of wide-ranging interests and talents.

He was also the cousin of Charles Darwin, the originator of the theory of evolution, with whom he often compared ideas and philosophy.   Galton was greatly influenced by Darwin, especially after he read his cousin's famous book, **On The Origin of Species**. After that, Galton devoted much of his time to studying related subjects.

Sir Francis Galton

[III] *Heritability is a statistical concept.* It describes *how much of the variation in each trait can be attributed to genetic variation.* As an example, the heritability of height is about 0.9. This means that 90% of the variance within the population measured for height can be accounted for by the variance of genes within the population.

A question of high interest to Galton was whether human abilities, such as intelligence, were hereditary. There was no history of such studies, so he had to invent his methodologies.

First, he conducted studies of several generations of families with eminent men. His hypothesis was that, if the traits that make a man eminent were heritable, then there should be more eminent men among their relatives than among the general population.

The results of these studies proved his hypothesis, and he described his findings in his book, **Hereditary Genius**, published in 1869.[15] Cousin Charles Darwin, found the effort to be of high interest and encouraged Francis to continue in his work.

Galton followed up this endeavor with a survey of twins raised in different environments. His objective was to understand the impact of nature vs. nurture on the twin's development. He published his results in an 1875 paper, "The History of Twins, as a Criterion of the Relative Powers of Nature and Nurture"[16].

His conclusion was that the *evidence favored nature* in the inheritance of abilities such as intelligence and other skills. With this approach, Galton anticipated the *modern field of Behavioral Genetics* which conducts numerous twin studies investigating "heritable" traits.

This interest in the heritability of positive, human traits *led Galton to invent the term "eugenics"* in his 1883 book, **Inquiries into Human Faculty, and its Development**.[17] He wrote in the Introduction:

> *"[This book's] intention is to touch on various topics more or less connected with that of **the cultivation of race**, or, as we might call it, with 'eugenic' questions, and to present the results of several of my own separate investigations."*

As far back as ancient Greece, people have practiced what *we now call eugenics. This is the philosophy that has the goal to improve the human race by encouraging people with positive and desirable traits to reproduce (positive eugenics), while discouraging reproduction among people with "undesirable qualities" (negative eugenics).*

As a result of Galton's interest in eugenics, in the early part of the twentieth century, *Galton publicized his ideas and started a popular eugenics movement in England. It then spread to other European countries and finally the United States.*

Based on today's standards, eugenics is viewed as a deplorable policy. But, in those times, the movement was propelled by the demographic chaos created by the higher levels of industrialization that were just developing in advanced European countries. *This led to the related immigration of workers to those countries looking for opportunities to improve their quality of life.*

Apparently, many of these immigrants were viewed by the host countries' populations as being "undesirables". *Therefore, the eugenics-inspired policies, intended to "purify the race", called for forced sterilizations of these undesirables, and restrictions on interracial marriage.*

In 1907, the *Eugenics Education Society* was founded in Great Britain by some leading intellectuals including the well-known writers H.G. Wells and Aldous Huxley. They were concerned about the threats to society posed by the immigration of "undesirables".

In 1909, Galton was made the Honorary President of the Society. The organization published the **Eugenics Review** which continued until 1968, when it changed its name to **The Journal of Biosocial Science,** and which continues to this day.

In the beginning of the movement, eugenicists struggled to find a viable means to measure differences in people that would allow them to sort the "good", or desirable, from the "bad"

They tried measuring the crania of school children, analyzed the facial asymmetries of criminals and sketched the toes of prostitutes, looking for telltale indications. *None of these approaches provided the satisfactory confirmations they needed to define their policies.*

Coincidently, extensive IQ testing began in the early part of the twentieth century among large segments of the population. Data emerged from the tests showing significant differences in IQ for "the feebleminded". *Intelligence testing then became the tool of choice for the eugenicists to create their positive and negative eugenics programs.*

Eugenicists advanced the notion that countries needed to conduct large scale intelligence testing to identify, segregate and sterilize the "feebleminded", defined to be "people with mental disabilities, and all "unfit persons" of low intelligence, character or ethnicity".[18]

Poster of English Eugenics Movement

The eugenics movement in the United States was led by Charles Davenport, a biologist, and Harry Laughlin, a public high school teacher. Davenport founded the *Eugenics Record Office* on Long Island, New York in 1910, a few years after a similar operation started in England. Laughlin became the first Director of the Office.

The staff of the office was charged with collecting data to identify eugenics opportunities. The information included family histories indicating undesirable physical and moral traits such as mental disability, dwarfism, promiscuity, and criminality.

Harry Laughlin, along with Madison Grant, and others, founded the *American Eugenics Society* (AES) in 1926 to promote eugenics education programs. AES, under its president, Henry F. Perkins, worked with the organizational predecessor of *Planned*

Parenthood, to use birth control to improve the quality of the U.S. population.

In 1927, there was even a U.S. Supreme Court ruling that supported the eugenics movement. The famous case was Buck vs Bell. The court upheld the involuntary sterilization of Carrie Buck, measured as an "imbecile" (IQ of 26 to 50). Chief Justice, Oliver Wendell Holmes, wrote the majority opinion. He said,

> *"It is better for all the world, if instead of waiting to execute degenerate offspring for crime, or to let them starve for their imbecility, society can prevent those who are manifestly unfit from continuing their kind....".*

This was a major victory for the eugenicists. Their position was now ratified by the Supreme Court of the United States.

Carrie Buck

So, according to the Supreme Court, from a legal perspective, *the world's interests were to be considered before the interests of the "degenerate and inhuman" individuals who were found to be deficient through an IQ examination.*

The implementation of compulsory sterilization programs under Buck vs. Bell resulted in over 65,000 forced sterilizations of low IQ people in thirty-three states from the 1920s through the 1970s.

Many sterilizations were done with the permission of their families who were concerned about the future of their disabled children. But others were performed by State governments who targeted specific groups for sterilization. Victims were often sterilized without their consent or knowledge.

In 2015, in a show of remorse over these unfortunate policies that were in place decades earlier, the U.S. Senate voted to compensate living victims of involuntary sterilization.

The Fall of Eugenics

The overall eugenic movement in the U.S. peaked in the 1920s and early 1930s. Americans started to understand the implications for the movement and rejected the idea.

While the eugenics movement died out in the U.S., this was not the case elsewhere in the world. During the 1930s, the Nazis took eugenics to a whole new level. One of Nazism's major objectives was the biological improvement of the German people by selective breeding for "Aryan" traits (positive eugenics). Their goal was to produce the master race of non-Jewish Caucasians with Nordic features.

To accomplish this objective, the race was to be cleansed of all peoples "unworthy of life" (negative eugenics programs). The first groups to be eliminated were those people in private and state-operated institutions, such as prisoners, dissidents, and people with congenital, cognitive, or physical disabilities. 400,000 people under Nazi authority were sterilized against their will, while up to 300,000 were killed in a mass murder campaign.

German propaganda poster promoting sterilization

"Who would want to be responsible for this?

Then, once World War II started in 1941, the Nazis brought their eugenics campaign to an even higher, more evil level. They launched the systematic extermination of all European Jews in one of history's worst genocides. Over the next five years, an estimated six million people were murdered in the six, major German extermination camps located in Poland, including Auschwitz and Treblinka.

Auschwitz Prison Camp

The genocide continued until the camps were liberated by U.S. troops in 1945, near the end of the war.

It was not until seventy-five years after Buck v. Bell, that the U.S. Supreme Court finally had an opportunity to modify its attitude toward the mentally inadequate. In Atkins v. Virginia (2002), the Court put itself in the defendant's shoes and upheld the right of the low IQ person to be held non-culpable due to cognitive impairments.

In the majority opinion written by Justice John Paul Stevens, the Court held that executions of mentally retarded criminals are "cruel and unusual punishments" prohibited by the Eighth Amendment. Thus, the Court's action humanized the person with mental retardation and raised the need to consider their rights as individuals.

Racial Segregation

Another bad idea that was ultimately supported by IQ testing was racial segregation. Starting in World War I (circa 1917) the Army made extensive use of the IQ testing of recruits. The Army exams were utilized to assign job classifications and place everyone in his most appropriate role. There were Alpha and Beta versions of these tests.

The Army Alpha test was for literate recruits and measured "verbal ability, numerical ability, ability to follow directions, and knowledge of information"

The Army Beta test was used with foreign speaking recruits or those who were illiterate. Beta consisted of seven components that measured such things as the recruits' abilities to trace a maze, do cube analysis, do pattern analysis, code digits with symbols, number check, complete pictures and perform geometric constructions.

The Army test results inadvertently contributed to the social injustice of segregation by identifying significant differences in IQ scores based on race. The results led to the military policy of segregating African Americans and placing them in the more menial, non-combat roles. These policies were still in effect as the US entered World War II.

It was not until late in the war that troop losses forced the military to place African Americans in jobs like infantry soldiers, pilots, medics, and officers. *The men performed well in these roles.* So, *the desegregation policy **continued** after the war, but only unofficially.*

Tuskegee Airmen WWII

In July 1948, racial segregation polices in the military came to a formal end. In that year, President Harry S. Truman issued an Executive Order that established equal treatment and opportunity for all recruits in the U.S. Military, regardless of race.

Nature vs. Nurture

Arthur Jensen and IQ Racial Differences

The idea of differentiating the influence of "nature vs. nurture" in human development dates to the time of Shakespeare in the early 17th century. The question is "how much of what we see in a person's life is due to the innate qualities of the individual versus the impact of their subjective experiences?"

The argument is not whether both factors play a role, they do. The issue is how much does each count in the differences we see in individuals.

In the times before genetics was well understood, it was difficult to determine the role of "nature" or heredity. The best information available was just the status and state of the immediate family.

Obviously, a key environmental factor playing a role in the measurement of a person's intelligence is the living situation provided by one's parents and siblings. After that, some of the other factors include education, socioeconomic status, health, and nutrition. In some situations, testing bias and the impact of minority status are also considered.

As the U.S. culture changed in the 1950s and 1960s to a more egalitarian view, scientists and philosophers balked at openly studying racial differences in intelligence. *Numerous psychologists sought to show that the test results were much more dependent on cultural and environmental factors than genetics. They became the dominant group in the study of intelligence during this period.*

But that did not stop some scientists from following their data and continuing to explore the idea of a strong genetic component in the differences in intelligence between individuals and groups.

Arthur Jensen, like Francis Galton, became strongly associated with the idea that intelligence differences by race were largely due to heredity. Jensen was an American psychologist and long-time professor of educational psychology at the University of California, Berkley (1958-1994).

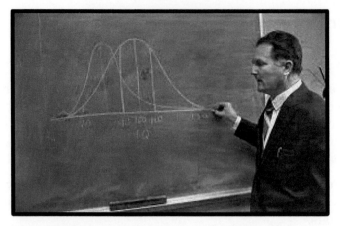

Professor Arthur Jensen, UC Berkley

Jensen did considerable research and testing of children and concluded that "g" is essentially an inherited trait. In 1969, he published his findings in the *Harvard Educational Review*. *"How Much Can We Boost IQ and Scholastic Achievement?"[19] was one of the most highly referenced and controversial articles in that journal of all time.*

The reason it was so controversial was because one of Jensen's conclusions was that the Johnson administration's *Head Start program* had failed. Head Start was part of Johnson's War on Poverty, designed to boost African American children's IQ scores. The program provided comprehensive , early childhood education, health, nutrition, and parent involvement services to low-income children and families.

Jensen concluded that the failure of Head Start was inevitable since more than 50% of the variance in IQ in the population studied was the result of genetic factors. Therefore, he said, attempts to increase intelligence by such a program as Head Start were doomed to fail.

Since low IQ was also associated with several other negative, social outcomes (e.g., drug use, crime, etc.), Jensen's finding about Head Start challenged most of the other Johnson programs aimed at remedying these problems. As a result, there was a *major, negative reaction against Jensen's work by many scientists with Progressive views.*

With his infamous article, Jensen's name became associated with his theory and the term, "Jensenism" became widely used. The Random House and Webster's Unabridged Dictionaries contain the following entry:

> **Jen-sen-ism** *(jen'se niz'em), n. the theory that an individual's IQ is largely due to heredity, including racial heritage. [1965-1970]; after Arthur R. Jensen (born 1923), U.S. educational psychologist, who proposed such a theory; see -ism]—Jen'sen-ist, Jen'senite', n., adj.*

In 1998, four years after retiring as a UC-Berkeley professor, Jensen published a book expanding on his research findings and conclusions, **The g Factor: The Science of Mental Ability**.[20] In Chapter 12 of the book, Jensen presents his data that suggest that IQ race differences are about 50% heritable.

Since then, the terms racist and white supremacist have become weapons of the Left, applied to anyone, such as Jensen, who dares to bring up racial differences. So, as it has been said, "there is no single accusation that will stop all conversation more quickly than to call someone a racist."

Not everyone agreed that Jensen was a racist. One such person was Frank Worrell, currently a U.C. Berkeley educational psychologist. In the 1980's, Worrell was a student at Berkeley and took a class from Jensen. Worrell was a native of Trinidad and is of African descent.

Professor Frank Worrell, UC Berkeley

When Worrell was considering taking Jensen's course, some of his friends advised him not to do so. They said Jensen was a racist.

But Worrell took the course anyway and found that the warning was unfounded. In fact, Worrell received the highest grade in Jensen's class. Worrell, speaking of Jensen, said, "He was a very nice guy," and "I never saw any sign of racism." [21]

What Worrell had no trouble recognizing was Jensen's devotion to science and data-driven research.

> *"I argued a lot, but his thing was 'Show me your data,'"* *Worrell said. "I know many of his ideas come across as racist to others, but he was interpreting the data as he saw it."*[22]

Jensen, himself, defended his positions as being scientific, and not racist when he said in an interview with the **London Times Higher Education Supplement** in 1996:

> *"The study of human differences cannot be racist."* *Comparing himself to anthropologists and medical researchers who study physical differences between racial groups, he said, "I'm simply doing the same thing with this trait, called general intelligence"*[23].

The Bell Curve Wars

The controversy over the genetic basis of intelligence and its impact on racial differences continued with the advent of another book on the subject. In 1994, exactly twenty-five years after Jensen's famous *Harvard Educational Review* article, the book, **The Bell Curve: Intelligence and Class Structure in American Life** was published by Richard J. Herrnstein and Charles Murray[24].

Herrnstein was a renowned professor of psychology at Harvard University. He had been the Department Chair between 1967 and 1971. Murray was a conservative scholar at the American Enterprise Institute, with an M.I.T. doctorate in Political Science.

The authors presented many insightful, *non-controversial* conclusions in their book. For example, *that human intelligence is derived from both inherited and environmental factors. And that variations in intelligence are better predictors of life's outcomes than an individual's parental, socioeconomic status.* Some of the outcomes they considered were financial income, job performance, birth out of wedlock and involvement in crime.

Charles Murray and Professor Richard Hernnstein,

Harvard University

But **The Bell Curve** was *a highly controversial work* because, like Arthur Jensen's writings, the authors *considered the connections between race and intelligence,* and then suggested social policies based on those connections.

The reactions to the ideas presented in the book were strong, both by supporters and critics. There was a flurry of articles and other books. **The Bell Curve** was termed, "The book event of the decade".

Herrnstein, unfortunately, died in September 1994, just as **The Bell Curve** was released. So, that left Murray to manage all the criticisms. Murray responded to the negative comments about genetics and race by pointing out that he and Herrnstein did not conclude that intelligence was determined entirely by genetics. He directed critics to a paragraph in the book that indicates that the authors believe that the difference in Black people's IQ scores is caused by both genetic and environmental factors. The paragraph reads:

"If the reader is now convinced that either the genetic or environmental explanation has won out to the exclusion of the other, we have not done a sufficiently good job of presenting one side or the other. It seems highly likely to us that both genes and the environment have something to do with racial differences. What might the mix be? We are

resolutely agnostic on that issue; as far as we can determine, the evidence does not justify an estimate."

The tone of **The Bell Curve** critics who were journalists is well represented by Joan Walsh, a columnist for **Salon.** She framed her criticisms as a response to Murray's defense, as illustrated in the above paragraph[25]:

"Let me grant this to Murray: That paragraph is indeed in his book. But the rest of its 800 pages are devoted to arguing that blacks and Latinos have lower IQ than Asians and whites (whites are inferior to Asians, by the way); that IQ is largely (though not exclusively) hereditary; that lower IQ means these groups are more likely to commit crime and drop out of school and have illegitimate (and lower IQ) babies and live-in poverty, and that there's not much to do to help those groups rise. In fact, Murray and Herrnstein argued, American welfare policies that provide aid to women with children inadvertently social engineer who has babies, and it is encouraging the wrong women.

"When you've spent an entire book arguing that blacks and Latinos have lower IQs, more out-of-wedlock babies and higher reliance on welfare, it's clear who "the wrong women" are. Oh, and the book also argued for limiting immigration, because unlike earlier waves of immigrants, todays are coming from countries with a lower national IQ. In what world are those arguments not racist?"

Many scholarly professionals, such as Stephen Jay Gould, the Harvard paleontologist and popular science writer, were also harsh in their criticisms of the professional work represented in the book. Gould's review was published in *The New Yorker.*

The Harvard scientist said that **The Bell Curve**:

"contains no new arguments and presents no compelling data to support its anachronistic social Darwinism. The authors omit facts, misuse statistical methods, and seem unwilling to admit the consequence of their own words."[26]

As a result of the major controversy over **The Bell Curve**, *The American Psychological Association* (APA) determined that *many participants in the debate made little effort to distinguish scientific issues from political ones.* One result was that there were serious misunderstandings about the research findings regarding the relationship of intelligence, genetics and environmental influences.

So, the APA formed a Task Force of eleven members to produce a "dispassionate survey of the state-of-the-art" (as of 1996) to make clear "what is scientifically established, what is in dispute and what is still unknown."[27]

Some of the Task Force key findings were:

- IQ tests do correlate with one another ("g" is a valid concept)

- IQ scores do correlate with various successful outcomes

  o Grades

  o Achievement tests

  o Various measures of job performance

- IQ scores had *negative* correlations with juvenile crime

  o Higher IQ score, less likely to commit crimes

- Genetics and environmental variables are both involved in the development of intelligence, *but genetics increases in importance with age*

  o Shared family environment is important for IQ scores in early childhood, but is *not* related in late adolescence

  o For a child to receive extremely poor or interrupted schooling is an environmental factor that is known to *negatively* affect IQ

- The Head Start intervention program produced initial IQ gains, but those gains disappeared by the end of elementary school

- Several biological factors, such as malnutrition, and exposure to toxic substances, resulted in lower IQ under some conditions

- *There is a long-standing, fifteen-point difference between the IQ scores of African Americans and White Americans*

  o The way the tests were formulated and administered *did not* contribute to the difference

  o Culturally based explanations for the difference *have not* been conclusively supported

  o *There is yet little empirical support for genetics as an explanation*

The APA review verified much of the research and reporting presented in **The Bell Curve.** But the bottom line of the document is that there is **no satisfactory explanation, either genetic nor environmental, for the measured differences in the means of the IQ bell curves of Black people, Whites and Asians.**

Since that report was made, there has been considerable progress in behavioral genetics and DNA research regarding the heritability of human intelligence, and we appear to be approaching a more definitive answer on the role of "nature" (see Section I, *The DNA Revolution*).

In 2014, twenty years after the publication of **The Bell Curve**, Charles Murray was interviewed.[28] He was asked how he responds to those who say his book tried to prove that Black people were genetically inferior to whites. He said, in part:

> *"The lesson, subsequently administered to James Watson of DNA fame, is that if you say it is likely that there is* **any genetic component** *to the black-white difference in test scores, the roof crashes in on you. On this score, the roof is about to crash in on those who insist on a purely environmental explanation of all sorts of ethnic differences, not just intelligence.*
>
> *Since the decoding of the genome, it has been securely established that race is not a social construct, evolution continued long after humans left Africa along different*

paths in different parts of the world, and recent evolution involves cognitive as well as physiological functioning."

*"We're not talking about another 20 years **before the purely environmental position is discredited,** but probably less than a decade. What happens when a linchpin of political correctness becomes scientifically untenable? It should be interesting to watch. I confess to a problem with schadenfreude."* [IV]

In less than five years, Murray got his wish. As shown in the section "Genome-Wide Polygenic Scores (GPS)", the role of nature in the nature vs. nurture argument in intelligence is clearer.

While nurture plays a strong role, the results of the numerous twin studies done by behavioral geneticists, and recent findings from studies looking for the genes related to intelligence, have *proven to everyone's satisfaction the equally powerful role also played by genetics in an individual's intelligence.*

The Rise and Fall of the Flynn Effect

Every fifteen or twenty years, the various IQ tests are revised by their publishers. Then a random sample of the population takes the new test, and their scores are used to normalize the results. A standard score of IQ 100 is defined as the median performance of the standardization sample. And all the scores are distributed in a normal, bell shaped, curve.

It has become the practice of publishers to have this new test population also take the previous IQ exam. For many years, the mean score of the new population taking the previous exam was significantly *above* that of the old population. One example is that British children's' average IQ scores rose 14 points from 1942 to 2008.[29] That is an increase of 2.1 points per decade.

This secular rise in IQ scores was found to hold in most countries where subsequent, norming tests were performed. The average

---

[IV] *Schadenfreude* - pleasure derived by someone from another person's misfortune.

rate of increase, when all the examples are considered, appears to be about three IQ points per decade.

James Flynn, now an Emeritus Professor of Political Studies at the University of Otago in New Zealand, documented and publicized the phenomena in his research work.[30] But it was Herrnstein and Murray, authors of The Bell Curve, who coined the term, The Flynn Effect.

Professor James Flynn

When researchers started to look more closely at what was happening, they found some interesting developments. In a longitudinal study, two large samples of Spanish children were measured for their IQ, with a 30-year gap between tests.

The Flynn effect was evident. The mean IQ scores on the two distributions increased by 9.7 points, or 3.2 points per decade.[31] The interesting finding, however, was that the gains were concentrated in the lower half of the population, while remaining negligible in the top half. *The gains gradually decreased as the IQ of the individuals increased.*

Many researchers became interested in the Flynn Effect and conducted studies that attempted to identify the causes for the IQ score increases. *There is, however, no definitive answer from this work.* But the most common explanations are that education and test taking skills have improved, or people had more stimulating environments, or they had better nutrition, or that there had been a decrease in the intensity of infectious diseases in developed countries. We will consider each possibility.

Our environment today is filled with many types of visual stimulations that did not exist forty or fifty years ago. Access to movies is greatly increased, as are television shows of all types, video games, and innumerable computer applications that are accessible on multiple devices including iPhones and iPads. *This has made the environment much more complex and stimulating to the minds of the users.* And this change may account for some of the increased IQ of the Flynn Effect.[32]

James Flynn, in his book, **What is Intelligence?** showed that environmental changes, such as those described above, means that, today, people are manipulating abstract concepts more than people who lived fifty years ago. Since a significant portion of an IQ exam deals with this ability, it can account for some of the improvement in test scores.[33]

Another hypothesis that was investigated by various researchers was that *improved nutrition* of the subject populations could account for some of the IQ increase.[34] Roberto Colom, of Universidad, Autónoma de Madrid, Spain, did the longitudinal study of the Spanish children in 2005. *He found that gains in IQ scores occurred predominantly at the low end of the IQ distribution.* This supported the improved nutrition hypothesis since it is the low-end part of the population where nutritional deprivation was most severe.

The *declining prevalence of infectious diseases* has been found to have an impact on the development of human brains, and thus IQ scores. The reason is that *brain development requires considerable metabolic resources.* Eppig, Fincher, and Thornhill argue that:

> *"From an energetics standpoint, a developing human will have difficulty building a brain and fighting off infectious diseases at the same time, as both are very metabolically costly tasks"* and that *"the Flynn effect may be caused in part by the decrease in the intensity of infectious diseases as nations develop."*[35]

While the IQ scores went up around the world for many years, it is still controversial whether "g" went up. Ulric Neisser, who led the American Psychological Association review of the state-of-the-art in intelligence studies in 1996, raised some red flags on the Flynn Effect in his own research.

Neisser noted, that, using the IQ values of 1997, the average IQ in the U.S. in 1932 would have been 80. So:

> *".... nearly one quarter (of the population) would have appeared to be (mentally) deficient." This was hardly the case. He concluded, "Test scores are certainly going up all over the world, but whether intelligence itself has risen remains controversial."*[36]

*However, **several studies done more recently found that the IQ scores of large populations flattened and started to decline during the 1990s.*** This is the reverse of the Flynn Effect. There are theories about the reasons for the decline, but it is still unexplained.

Separate tests of military conscripts in Scandinavian countries illustrated the falloff. Norwegian conscripts were given intelligence tests from the 1950s through 2002. The researchers, J. Sundet, et al, found that there was an increase in IQ scores up to the 1990s, *but then the scores flattened and declined thereafter.*[37]

Similarly, other researchers looked at the IQ test results for Danish conscripts. Teasdale and Owen found that, between 1959 and 1998, the young men's IQ gains first rose and then flattened out. *But between 1998 and 2004, only a six-year period, the IQ scores declined by more than 1 point.*[38]

The authors speculated that the decline might be due to changes in the Danish educational system. But they also found that draftees, who were first- or second-generation immigrants, scored below the average of the group. So, the authors speculated that the rising proportion of immigrants, or their immediate descendants in the population of cadets, were the cause of the decline.

James Flynn, as reported by Richard Gray, found similar results in his research in the U.K. He observed that tests of 14-year-olds, done in 1980 and again in 2008, showed more than two points decline over the period. The upper half of the population showed the most decline. Flynn attributed the decline to a stagnation or dumbing down of the youth culture in the UK.[39]

While looking at the declining IQ scores in several European countries, researchers Lynn and Harvey considered the fact that there was significant, recent immigration from countries with lower average, national IQs.[40] They expect the down trend in IQ scores to continue since, they speculate, there is a limit to how much environmental factors can improve intelligence.

Lynn and Harvey also found that, over the last century, *there was a negative correlation between fertility and intelligence.* The groups with the lowest IQs had the most children. Thus, they concluded that the overall IQ scores in the country would be diminished over time.

The DNA Revolution

It was in 1953 that James Watson and Francis Crick made the widely publicized discovery of *the double helix form of the DNA (deoxyribonucleic acid) molecule.*

**James Watson (left) and Francis Crick
with their DNA model in 1953**

Watson and Crick's breakthrough was one of the most important discoveries of all time because these scientists found that the *double-helix structure contains* the *complete genetic information needed* for the development, functioning, growth, and reproduction of all living organisms. In essence, t*he DNA instructions are what make each species unique.*

Every cell in a person's body holds exactly the same DNA material, and it is unique to that individual. That is why a researcher can obtain your DNA from any of your cells, like a flake of skin or a strand of hair.

*It amazing to think that **99.9%** of an individual's DNA is the same as every other human being. **It is only the remaining 0.1% that makes us different from one another**.*

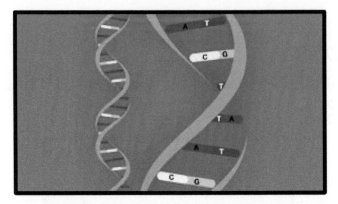

The Double Helix Molecule

To understand how DNA works, and how it affects our intelligence, we need to grasp a *few of the basics of genetics.*

DNA is a long *polymer,* or chain of organic molecules, called *nucleotides.* They are held together by *a backbone* made of sugars and phosphate groups. ***This backbone represents the sides in the twisted, rope ladder description of a DNA molecule as shown in the diagram above.***

Attached to each sugar molecule in the backbone is *one of four, nitrogen-containing, organic bases.* They are adenine (A), thymine (T), cytosine (C), and guanine (G).

As shown in the figure of the Double Helix molecule, adenine only pairs with thymine, (A-T)while cytosine only pairs with guanine (C-G). *These pairs make up the rungs of the DNA twisted-rope ladder.*

A gene is composed of a small section of the ladder, as shown on the right of the diagram below.

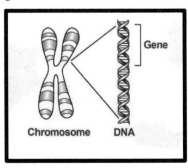

Genes found in Chromosome

Contained in the gene is the genetic code for a particular trait, or phenotype. The code is composed of a specific combination of the A-T and C-G organic bases in the rungs. It is this combination of bases that tells the cells *which protein to make that will determine the particular trait associated with that gene.*

Chromosomes are structures which hold the genes. They are thread-like arrangements, shown as an X shape in the diagram above. *Each chromosome is made up of long chains of DNA molecules, tightly coiled many times around proteins called histones.*

The histones provide support for the chromosome arrangement as illustrated in the diagram. *The chromosomes, in turn, are all enclosed within the nucleus of the cell.*

A single chromosome has hundreds to thousands of genes. *And every cell in the human body contains twenty-three **pairs** of chromosomes.* One set of twenty-three from the male, and the other twenty-three from the female.

*The term **genome** refers to all of the DNA contained in the cell. The human genome contains about 21,000 genes made up of 3 billion, base pairs **that spell out the instructions for making and maintaining a human being**.*

The genetic information, stored in the sequence of A-T and C-G base pairs that make up the rungs of the ladder, *can be arranged in huge variety of sequences and represent a vast, potential store of information.*

A gene uses this information to encode the sequence of amino acids that are the constituents of the proteins required for its particular trait. ***It is this process that determines how the genetic code is applied to make us who we are.***

A person's *traits*, like height and eye color, are technically called *phenotypes*. The phenotype's characteristics are determined by the genes. *In most cases, there are multiple genes involved in determining a particular trait.*

To complicate matters, *for some traits, such as intelligence, each gene has an exceedingly small effect.* In fact, more than one thousand genes might be involved in determining the exact characteristic. *As shown later in this section, this is the case for intelligence.*

Behavioral Genetics

Sir Francis Galton was an early researcher into the heredity of human intelligence (see section, *Rise and Fall of Eugenics*).   In fact, he was the first researcher to use twins to study the heritability of intelligence. He published his results in an 1875 paper, *"The History of Twins, as a Criterion of the Relative Powers of Nature and Nurture"* [41].

His conclusion was that *the evidence favored nature* in the inheritance of abilities such as intelligence and other skills. With this twin study approach, Galton anticipated the emergence of the modern field of *behavioral genetics* which still conducts numerous twin studies investigating heritable traits.

Since Behavioral Genetics is about determining the "heritability" of genetic traits, it is important to understand the technical meaning of the term.

Heritability is a statistical concept. It describes how much of the variation in each trait can be attributed to genetic variation. As an example, the heritability of height is about 0.9. This means that 90% of the variance within the population measured for height can be accounted for by the variance of the relevant genes within the population.

We need to keep in mind that there has been a non-scientific barrier to conducting research into the heritability of human intelligence. As a result, behavioral genetics research efforts have been greatly hampered.

After the Civil Rights movement succeeded in the 1960s, there was a major change in the culture of Western countries, away from racial discrimination and toward egalitarianism. This change had an unintended consequence in the way research was conducted into the genetic basis of human intelligence.

Many scholars came to view academic interest in the racial differences in intelligence as inherently, morally suspect, or even racist. And this attitude had a major impact on those researchers who dared to publish their research into this area, such as Charles

Murray, and Richard Hernnstein (**The Bell Curve**) and Arthur Jensen.

In fact, it got to the point that even research into *just the idea of the heritability of intelligence* became suspect. But, in spite of this obstacle, the behavioral geneticists continued to progress in their efforts.

One such researcher is Robert Plomin, an American psychologist and geneticist who has focused his career on behavioral genetics. He has been working in the field since 1974. In 1994, he moved to London to join the Institute of Psychiatry of King's College.

He has published widely in journals of *behavioral genetics*, especially regarding studies of the *heritability of intelligence*. His papers on the subject have been cited by many other researchers.

Professor Robert Plomin

Throughout his career, Plomin conducted numerous studies involving twins, both identical and fraternal. With the recent revolution in genetics technology, he has complemented the twin study approach with *Genome-Wide Association Studies (GWAS) (see the GWAS section below).*

In 2018, Plomin published an important book through the MIT Press in Cambridge, Massachusetts. It is **Blueprint: how DNA makes us who we are.** In the *Afterword* of the paperback version published in 2019, Plomin discussed his fears before the original, hardcover printing of **Blueprint.** He explains the impact that the "politically correct" environment had on his intelligence heritability research,

"I was worried about Blueprint's reception for the same reasons I waited thirty years to write the book. As described in the Prologue, it was dangerous back then, professionally, and sometimes personally, even to raise the possibility of genetic influence on who we are as individuals – our personality, mental health and illness and mental ability and disability. I thought the zeitgeist had been shifting toward genetics. One sign was the gradual acceptance in psychology and society of the evidence pointing to the importance of genetics. A second sign was the huge impact of the DNA revolution, beginning with the sequencing of the human genome in 2003."

After publication of **Blueprint**, Plomin *did* have to deal with significant criticisms from media pundits along the line of political correctness. But *his treatment was mild compared to that of Jensen, Murray and Hernnstein.* Plomin was able to respond to all the critics questions and, *in the end, the book was well received by the Public.*

Plomin, and a number of other behavioral geneticists, have made a major, surprising, and counterintuitive finding with regard to the heritability of intelligence.[42] The heritability of this trait has consistently, over three decades of research, *been found to increase linearly throughout the life of an individual.*

This was the case in longitudinal as well as cross-sectional analyses, and in adoption as well as twin studies.[43] *It is called the Wilson Effect, after Ronald Wilson, who, in 1978, was the first researcher to note the phenomena.*

In particular, the research showed the heritability of intelligence increases significantly from 41% in childhood (age 9) to 55% in adolescence (age 12) and to 66% in young adulthood (age 17)). Some evidence suggests that heritability can increase to as much as 80%, in later adulthood.

What this finding means is that, as a person ages, more and more of their measurable intelligence is determined by their genes, and less by the environment.

Increasing heritability for intelligence is interesting and counterintuitive because other domains, such as personality, do not show systematic changes in heritability during development.

So, why does heritability of intelligence increase throughout development? Plomin, et al, believe that it could be *due to new, additional genetic influences coming into play as the person ages.* This is a genetic process called *innovation.* And there is some evidence for it as *the research hunt progresses to identify all the gene's responsible for intelligence.*

Genome-Wide Association Studies of Intelligence

There is a great deal of research interest in developing a method to *identify which gene, or group of genes, are associated with a particular trait such as intelligence.*

The researchers started by using procedures where they focused on one candidate gene at a time. It was very frustrating for them. The reason was that most of those gene candidates turned out *not* to be responsible for the trait. Moreover, there are about 21,000 genes in the human genome, making the task monumental.

So, the *scientists needed a new, more efficient approach.* The answer was the development of *Genome-Wide Association Studies (GWAS).*

This multi-step technology (see Figure below) is used to *narrow down the search for the **genes associated with intelligence to the area of the genome where the target genes reside.***

To do a GWAS, you *identify a population of humans who has the trait you want to study, say IQ greater than 120, and another group of similar people who do not have the trait.* Then DNA is obtained from each participant, using a blood sample or a swab inside the mouth.

Each person's genome is then obtained from their sample using a high-throughput, DNA purification system. The DNA samples are then placed on tiny chips and scanned with a computer-automated, Machine Vision system.

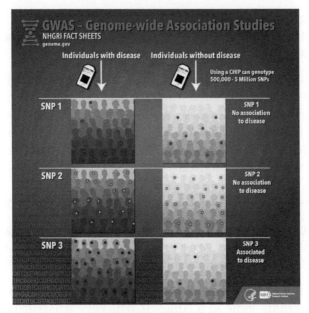

GWAS Study Design

The vision system identifies selected markers of *genetic variation in the sample.* These are called S*ingle Nucleotide Polymorphisms,* or SNPs (snips).

For example, one person in the high IQ group might have the nucleotide triplet TAC, whereas another person in the low IQ group might have TCC. The hypothesis is that *this identified variation may contribute to differences in intelligence between the two individuals.*

If particular genetic variations are found to be statistically more frequent in subjects with the intelligence trait compared to people without it, these genetic variations are *determined to be "associated" with the trait.*

These associated variations, therefore, **point to the region of the genome where the trait resides. This greatly narrows down the area of the search.**

The next step in the genetic identification process is **sequencing** *DNA base-pairs* in the *detected region of the genome.* The sequencing determines the order of the four bases, A-T and C-G, which make up the rungs of the DNA molecule.

The order tells scientists the genetic information that is contained in that DNA segment. And *allows them to find the exact genetic change involved in the intelligence trait.*

A key, government-funded organization for genome study is the **National Human Genome Research Institute,** located in Bethesda, Maryland. *They are leaders in developing genetic research technology. Here is what they say about the direction of technology for DNA sequencing:*

> *"Another new technology in development entails the use of nanopores to sequence DNA. Nanopore-based DNA sequencing involves threading single DNA strands through extremely tiny pores in a membrane. DNA bases are read, one at a time, as they squeeze through the nanopore. The bases are identified by measuring differences in their effect on ions and electrical current flowing through the pore.*
>
> *Using nanopores to sequence DNA offers many potential advantages over current methods. The goal is for sequencing to cost less and be done faster. Unlike sequencing methods currently in use, nanopore DNA sequencing means researchers can study the same molecule over and over again."*

It is new technologies, such as nanopore-based DNA sequencing, that *will lead us to a more-clear understanding of the role that certain genes play in the heritability of intelligence.*

Genome-Wide Polygenic Scores

One of the complicating factors in the search for the full range of "intelligence genes" comes from a *key finding in the early studies that were conducted.* **It is that the heritability of intelligence is caused by exceedingly small effects from many, many genes**.

Although the small effect size of individual DNA variants make it more difficult for neurocognitive research, it has been found that *polygenic scores* can be created from the GWAS that *aggregate* the effects of the DNA variants.

In fact, the *key output of the statistical model that analyzes the GWAS data is called* a ***genome-wide polygenic score (GPS)***. It is the *summary of the effects of the many genetic variants* found in the GWAS that relate to an individual's intelligence.[V] Some of the effects are positive, and add to a person's cognitive abilities, while others detract from it.

In early studies applying GWAS technology to search for intelligence genes, population sample sizes tended to be a few thousands of individuals. This size of population was thought to be adequate to obtain statistically significant results. *But, as it turned out, the outcomes were not useful because of the exceedingly small, individual gene effects.*

The gene search started to bear fruit in 2017 *when an innovative approach was tried.* Researchers from a Dutch university *used meta-analysis with a* combined population sample size of 78,308 to successfully identify inherited, genome sequence differences.[44]

They found twenty-two genes that had an impact on intelligence. *While not nearly complete, it represented a good beginning.*

This approach worked because *meta-analysis is a research process that systematically merges the findings of multiple, independent studies, using statistical methods to calculate an overall or 'absolute' effect.*[45]

Meta-analysis does *not* simply pool data from smaller studies to achieve a larger sample size. Analysts use well recognized, systematic, statistical methods to account for differences in sample size and variability (heterogeneity) in the studies.[46]

Larger, combined sample size studies, on the order of 270,000, conducted subsequently to the Dutch study, have found genetic variants that account for 20% of the approximate 50% heritability of intelligence.[47] *206 genomic loci were identified implicating 1,041 genes.*[48]

[V] **The GPS** is calculated as the *weighted sum of the alleles* by their effect size. These alleles are the multiple forms of the intelligence-associated genes found in the GWAS.

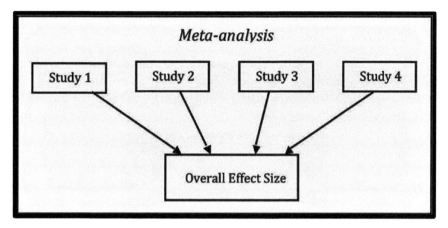

There is, however, still considerable "missing heritability". This is the *large gap between the known heritability of intelligence, as determined from twin studies conducted by Robert Plomin and others, and the variance that is explained by associations with specific DNA variants from the GWAS studies.*

Because of the small effects from so many genes, the size of the population sample used in a meta-analysis must be exceedingly large to find all the intelligence related variants. *It is estimated that the sample size will have to approach one million individuals to be adequate.* Indeed, this is the current direction of research.[49]

It should be noted that the GPS is a "causal predictor". That is, nothing in our brain or the environment can change the differences in the DNA sequence that we inherited. So, according to Robert Plomin, *an IQ GPS, derived from a DNA sample taken in adulthood, will predict that adult's intelligence. Likewise, DNA that was obtained at birth will have the same results as the adult sample.*

Thus, the GPS reflects an individual's estimated genetic predisposition for intelligence and can be used as a predictor for that trait.[50] This leads to the ***controversial idea (see below)*** that the GPS can be used to provide a measure of an individual's intelligence *without having to take an exam.*[51]

To turn a personal GPS into a DNA IQ test, you simply add up all the pluses and minuses for the intelligence related variants you find in the person's genome. Although some commercial services such as GenePlaza and DNA Land, are offering to tell you your inherited IQ potential from a sample of your DNA, the *results are of low quality and virtually useless.*

Until the missing heritability is found, the commercially available IQ GPS will continue to be poor. *But Plomin is optimistic that the technology will eventually allow meaningful, inexpensive estimates of a person's intelligence.*[52]

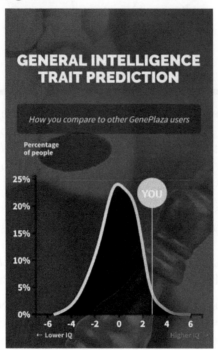

Sample output of GenePlaza Intelligence prediction

The question is, will such knowledge be welcome? Many people are already concerned about the implications. Droves of individuals are currently providing their DNA samples to commercial analysis companies and will continue to do so.

When the companies believe that the technology is ripe, the IQ GPS will be done as part of the service, and the information will be added to the firm's large data bases on individuals. *The concern is, how that data will ultimately be used.*

Similar to the history of eugenics (discussed in the section on *IQ Measurement Controversy*), having IQ data available could present serious problems for citizens. Think about the Nazis. It could be tempting for future people in authority to use readily available IQ GPSs for making decisions about the individuals in the data base.

Other troublesome issues include, "will schools use the information to separate children into classes differentiated by their IQ GPS?" Will your IQ GPS limit your opportunities for employment? Will government entities use the information in making policy decisions about the allocation of resources to low IQ individuals?

There is no doubt that highly accurate, IQ GPS will soon be with us. It is inevitable that *all these issues will emerge and should be the subjects of future discussions.*

The major question is, *"will the discussions lead to effective regulations that protect the rights of individuals?"* **Humans have a spotty record in that regard.**

Manipulating intelligence: CRISPR-Cas9

CRISPR (pronounced "crisper") technology is going to have a major impact on the world in the most profound ways. The letters stand for "Clusters of Regularly Interspaced, Short Palindromic Repeats" that are found in a DNA sequence. CRISPR *is a cheap, easy, and effective means to make **heritab**le changes in human embryos.*

*CRISPR was adapted from a naturally occurring, **genome-editing system** present in bacteria.* The bacteria uses the editing system as an innate defense mechanism to foil invading attacks on the bacteria by viruses or other foreign bodies.

Essentially, *CRISPR is a specialized stretch of DNA.* The CRISPR *acts as a homing device* that guides Cas9, another protein molecule that *acts as a set of scissors.*

Together, these two molecules can repair, disable, or insert something new into the target gene. **This means researchers can easily alter DNA sequences and, thereby, modify genes' functions.**

One of the pioneers in CRISPR technology is *Jennifer Doudna*, a professor at the University of California, Berkley. She is a biochemist and was co-awarded the 2020 Nobel Prize in

Chemistry for this work. She shared the prize with *Emmanuelle Charpentier*, now at the Max Planck Unit for the Science of Pathogens in Berlin.

Jennifer Doudna (left) and Emmanuelle Charpentier

The technology has *many positive, potential applications. For example,* correcting genetic defects, as well as treating and preventing diseases.

Once all the genes involved in human intelligence are known, it would be tempting to use CRISPR-Cas9 technology to increase a baby's cognitive abilities. *But there could be unintended consequences for the changes made to the baby's DNA.*

One possibility for that happening is that the CRISPR system could operate on some of the baby's genes that were *not* intended to be deleted or altered. This can happen because the CRISPR molecule, while looking around the genome for its target, *could find another, unintended gene that is, however, also a match.*

There have been such accidents in the past. In 2002, a little boy developed leukemia after a gene he received landed at a spot in his DNA that activated a cancer-causing gene.

There is hope, however. *Scientists have made "tremendous progress" in minimizing CRISPR's off-target effects*, said Dr. J. Keith Joung of *Massachusetts General Hospital.* It is a complex and difficult problem, but much effort is going into the research.

"And, someday, all could go well and CRISPR technology will provide a cure for cancer."[53]

The Neuroscience of Human Intelligence

Neuroanatomy: The Structure/Functions of the Brain

Introduction

The human brain is a wondrous "machine". How it operates makes you the unique individual that you are. Your vast store house of long-term memories, your deep and complex, constant learning about the world around you, and the ongoing and profound processing of all the added information coming in through your senses, continuously changes you. *Scientists call it neuroplasticity.*

While the brain is the control center, the nervous system is the body's inner communication system. It is made up of the body's many nerve cells. The nerve cells take in information through the body's senses: touch, taste, smell, sight, and sound. The brain interprets these sensory cues to understand what is going on outside and inside the body. This allows a person to use their body to interact with their surrounding environment and control their body's voluntary functions.

The human nervous system is divided into two parts. They are distinguished by their location in the body and include the central nervous system and the peripheral nervous system.

The central nervous system is located in the skull and vertebral canal of the spine. It includes the nerves in the brain and spinal cord. All remaining nerves in other parts of the body are part of the peripheral nervous system.

Neuroplasticity is the ability of the human brain to re-wire itself. As new information is acquired, or if you are unfortunate enough to have some type of injury, the brain can form and reorganize its "wiring" to accommodate the changes. It is this property that allows us to continue developing ourselves from infancy to adulthood.

Before we look at the brain's relation to your level of intelligence, we need to understand the structure of the brain. This is called *neuroanatomy.* Then, to comprehend how it generates our degree of intelligence, we need to appreciate how each part functions. This is the study of *neuroscience.*

The Structure of the Brain

The figure below is an anatomical diagram of the right hemisphere of the human brain. The brain is symmetric, so the left hemisphere is the same structure as the right, although the two hemispheres have different, functional specialties.

The first thing to notice is that the brain is divided into three major structures (see the diagram below): 1. the cerebrum, 2. the cerebellum, and the 3. brain stem.

The largest part is the cerebrum. It is on top and is shown with its four lobes noted. The *cerebrum's surface,* **called the cortex,** is grey and is covered with many bumps (gyri) and grooves (sulci) that result in its folded-looking structure.

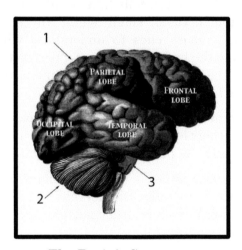

The Brain's Structure

This folding of the brain increases its surface area and enables *more cerebral cortex matter* to fit inside the skull.

This feature is important for intelligence since the larger the cerebral cortex, the greater the number of neurons that are available. Now, neurons are the basic, specialized cell of the brain. Therefore, the more neurons available, the more, higher level, mental processing that can be performed.

More about this phenomenon in a later section of this chapter.

And each hemisphere has four lobes as shown in the diagram above. They are:

1. **Frontal lobe**: front-most lobe, found near the forehead

2. **Parietal lobe**: behind the frontal lobe, near the top of the head

3. **Temporal lobe**: below the parietal lobe, near the bottom of the head around the ears

4. **Occipital lobe:** back-most lobe

The *cerebellum* (the narrow-striped structure on the bottom) is found underneath the occipital and temporal lobes. It looks different than the other structures, without any of the major bumps or grooves.

Finally, the *brain stem* is on the very bottom. It touches the temporal lobe and is surrounded by the cerebellum.

While the two hemispheres are symmetric physically, their functions are quite different as shown on the next page.

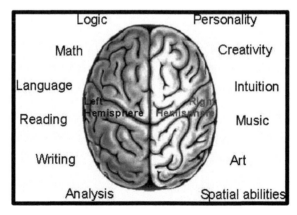

Functions of the two hemispheres

First, the *left hemisphere controls the right side of the body,* and the *right hemisphere controls the left side of the body.* The left hemisphere is responsible for logic and linear thinking. While the right hemisphere is the center for creative, non-linear activities.

Contained in the latitudinal fissure between the two hemispheres is the *corpus collosum*. It is a wide, thick tract of nerves that *connects the cerebral hemispheres and allows them to communicate with each other*. It is the largest, white matter structure in the brain and is about 10 centimeters, or almost four inches, in length.

The Functions of Brain Areas

We know from years of detailed, scientific studies which areas of the brain are involved in various activities. Historically, these studies were done by observing the capabilities lost by a patient with brain damage. In more recent times, that approach is supplemented using sophisticated technologies like *functional magnetic resonance imaging* (fMRI). This technique provides much more detailed and thorough information. The technology is described in the section, "The Centers of Intelligence in the Brain".

While one part of the brain may be the focal point of activity as found in the brain damage studies, fMRI shows that most actions use many sections of the brain at the same time. This was a major finding of this research. The figure below illustrates the *primary* brain locations for various actions or activities.

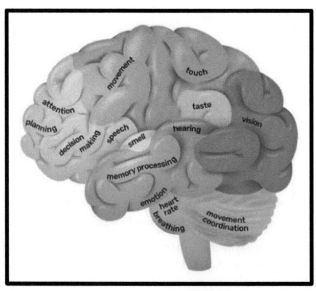

As shown in the diagram, the *frontal lobe* is generally where *higher-level thinking* and executive functions occur. It is part of the neocortex, the most recently evolved section of the brain. This is the outer layer of neural tissue on the surface of the cerebrum.

The major activities of the frontal lobe include focusing attention on a particular task, multi-tasking, planning, reasoning, problem solving, risk assessment and decision making. It is also the area of the brain where the complexities of social interactions are managed.

The *parietal lobe* is behind the frontal lobe. This area is responsible for sensations like touch, smell, hearing, taste, temperature and pain. And then it is this lobe where integrating the sensory information is done to form a single perception.

This part of the brain also plays a significant role in understanding numbers and their relations. Additionally, the *posterior* parietal cortex assists in coordinating movement. Finally, the small lobes on the parietal lobe, called lobules, are the primary areas for sensing body or spatial awareness.

The *temporal lobe, below the parietal lobe,* is also involved in processing sensory information. This activity is particularly important for hearing, recognizing language, and forming memories.

The temporal lobe also contains the primary *auditory cortex,* which receives auditory data from the ears, and processes that information so that we understand what we are hearing, whether it is words or environmental sounds.

The *medial* (closer to the middle of the brain) part of the *temporal lobe* contains a sub-structure called the *hippocampus. It is a region of the brain important for memory processing, learning and emotions.*

Parts of the temporal lobe are also involved in visual processing of complex stimuli such as facial features and object perceptions and recognition.

The *occipital lobe,* located in the rear of the head, is the visual management center of the brain, as shown in the diagram. In particular, the *primary visual cortex* receives data from the eyes.

Then, this information is relayed to several secondary visual processing areas, which interpret depth, distance, location, and the identity of seen objects. It is specialized for processing information about static and moving objects and is also excellent at pattern recognition. Additional visual tasks in this area of the brain include visuospatial processing, color differentiation, and motion perception.

The *cerebellum* is the second-oldest part of the brain, in terms of evolution, and is located at its base. About the size of a baseball, it extends around the brainstem. It plays a vital role in virtually all physical movements such as walking, driving a car, or throwing a ball. Beyond these gross movements, it also controls our fine motor skills, such as the subtler movements done by the fingers and hands. In addition, it assists with eye movements and vision. Altogether, it helps the body to maintain its posture, equilibrium, and balance.

Finally, the cylindrical structure at the base of the brain is the *brain stem*. It is the oldest, evolutionary part of the brain and connects the brain to the spinal cord. In this role, it carries all the information from the brain through the nervous system to the rest of the body, and information from the body back to the brain.

All the vital, autonomic, bodily functions are controlled by the brain stem. These include breathing, swallowing, digesting food, heart rate, and blood circulation. Furthermore, it also controls your involuntary muscles — the ones that work automatically, without you even thinking about it.

Brain Size and Intelligence

Since the 1830s, scientists have speculated that people with larger brains are more intelligent. As a result, many studies were conducted over the years to check the validity of this hypothesis. Before the advent of the MRI, most of these studies were done on *post-mortem brains.*

The evidence from these studies was only of limited value. First, there were few brains available for such research. Second, human brains tend to shrink after death and that introduces another variable.

The first, medically useful, MRI scanner was developed in 1980. But it was not until 1991 when a study was conducted by Psychologist Lee Willerman, et al, from the University of Texas, Austin, that *compared brain size, as determined by an MRI image, to the individual's performance on a standard intelligence test instrument.*[54]

The authors had a sample of forty college students. The students' brain sizes were measured with an MRI image. Then the students took four subtests of the Wechsler Adult Intelligence Scale-Revised (WAIS-R) including vocabulary, similarities, block design, and picture completion.

When the results were compared, the *researchers found that individuals with higher IQs had a larger brain size.* Factors such as body size and socioeconomic background were statistically controlled. *The researchers' conclusion was that human brain size is, in fact, positively correlated with performance on intelligence tests.*

Many other, similar studies were completed over the following years with comparable results. Then in 2005, Michael McDaniel of Virginia Commonwealth University conducted a *meta-analysis*[VI] of the relationship between in vivo brain volume and intelligence.[55]

[VI] **Meta-analysis** is a research process that systematically merges the findings of multiple, independent studies, using statistical methods to calculate an overall or 'absolute' effect.[VI]

McDaniel's analysis was based on thirty-seven sample populations with a total of 1,530 people. *The researchers found a positive correlation of larger brain size with higher IQ.* The correlation was higher for females than males. *For all ages and sex groups, however, it was clear that brain volume was positively correlated with intelligence.*

In the next section of the book, **Neurons: Building Blocks of the Brain,** *Role of Neurons in Intelligence,* we see *one reason why more intelligent people have larger brains.* It is that they have *larger,* pyramidal neurons, found in their cerebral cortex. along with dense, dendritic trees. In particular, these pyramidal neurons were found to increase the thickness of the cerebral cortex and therefore increase the brain's volume.

Neurons: Building Blocks of the Brain

Introduction

Now we know that the human brain has about 86 billion neurons on average,[56] thanks to the work of Suzana Herculano-Houzel of Vanderbilt University. She developed a unique methodology that she called "brain soup".

Her technique is a fast, accurate, reliable, and inexpensive way to count the number of neurons in brain matter. Since it is known that there is a direct correlation of intelligence with the number of neurons in the cortex, we can use her method to compare the intelligence levels of most animals.

We will look at her work in more detail in the chapter on *Animal Intelligence.*

Understanding how billions of neurons work together to produce the marvelous functions of the brain is the task of Neuroscientists. Their goal is to *map the human brain at the mechanistic level.*

To accomplish this goal, the scientists use a range of tools and skills to get at their difficult job. These include cellular and molecular biology, anatomy and physiology, human behavior, and cognition, as well as other disciplines.

As their contribution to this complex task, *The National Institutes of Health* (NIH) launched, *The Human Connectome Project. The* **Connectome** *is a comprehensive map of neural connections in the brain and the rest of the nervous system.*

Essentially, the Connectome is the nervous system's *"wiring diagram."* NIH's first Connectome-related project was launched in 2009. Now, there are dozens of studies underway looking at the development of the human brain across several age groups.

One technique that the NIH developed for this project was *a new design for an MRI scanner.* It has the strongest, magnetic gradients ever built for an in vivo, human MRI.

The purpose of the instrument is to map, as completely as possible, the *macroscopic, structural connections of the in vivo, healthy, adult human brain using diffusion tensor tractography*[VII].

This is a 3D modeling technique used to visually represent nerve tracts from the data collected by the MRI. The results are presented in two- and three-dimensional images called *tractograms*.

More specifically, *the Connectome will map out all the networks of cell-to-cell communication.* These are the brain circuits that process all thoughts, feelings, and behaviors.

It needs to be noted that these connections grow and change over time. This is a critical function of the brain called neuroplasticity and underlies all learning.

A thorough understanding of the anatomy of the neuron, the building block of the brain and the rest of the nervous system, is the essential, first step to comprehending brain function.

The Structure of the Neuron

The central nervous system, which includes the brain and spinal cord, is made up of two basic types of cells: neurons and glia. Neurons are the key players in the work of the brain. The glia cells play a supporting role.

As shown in the Figure below, neurons have three basic parts. First *a cell body, called the soma, containing the nucleus.* Then, as shown, there and two types of extensions from the cell body, dendrites, and axons.

The dendrites and the axon extend out from the soma and are the means for interneuronal communication. The dendrites *job is to* **receive** *electrochemical* **inputs** *from other neurons and transmit them to the cell body.*

[VII] **Diffusion tensor imaging tractography**, or DTI tractography, is an MRI (magnetic resonance imaging) technique that measures the rate of water diffusion between cells to understand and create a map of the body's internal structures; it is most commonly used to provide imaging of the brain.

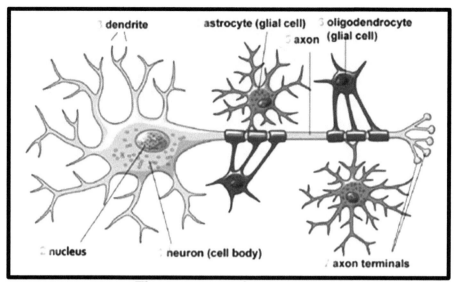

The structure of a neuron

On the right is an *axon* which has branched terminals on its end. *Its function is to **transmit** electrochemical **outputs** from the cell body* to other, upstream neurons.

Synapses are the junctions between two neurons. The synapses are the key to the brain's functions. They consist of tiny gaps across which impulses, acting as data, pass through by diffusion of an electrochemical neurotransmitter.

In other words, the *dendrites are branched, extensions of the cell body.* They receive and conduct to the cell body the electrical stimulation, or data, which comes across a synapse from the axons of other neurons.

Similarly, the axon has terminals on its end. It carries nerve impulses away from the cell body and through a synapse to the dendrites of another neuron.

The synapses are located at various points throughout the *dendritic tree,* or the branching out of the nerve fibers at the end of the dendrites. Each neuron can form thousands of synaptic connections with other neurons. So, *there are over 100 trillion such connections in the average brain.*

This system, as described above, makes up the interneuronal communication scheme. And it is responsible for the reception, processing, and communication of information throughout the brain and other parts of the body.

Also shown in the Figure displaying the structure of the neuron, are glial cells. There are four of these cells shown attached to the axon, *two astrocytes and two oligodendrocytes.*

The glial cells are *non-neuronal* cells that *do not produce electrical impulses.* Instead, they provide support and protection for the neurons. To do this, they have multiple tasks. First, they maintain homeostasis, or a stable equilibrium of the cell's functions. They also form myelin, the protective sheath around the axon.

The astrocytes specific functions include providing nutrients to the nervous tissue, maintaining extracellular ion balance, and playing a role in the repair and scarring process of the brain and spinal cord.

Oligodendrocytes create the myelin sheath around the axon. Myelin is a lipid-rich (fatty) substance that acts as an insulator. The *sheath tends to increase the rate* at which electrical impulses (called action potentials) are passed along the axon. The myelinated axon can be likened to an electrical wire (the axon) with insulating material (myelin) around it.

A single oligodendrocyte can extend its processes to fifty axons, wrapping approximately one micrometer of myelin sheath around each axon.

In summary, the neuron is the basic building block of the nervous system. It is an amazingly well-designed machine that performs a profound role in our lives, and helps to make us who we are, with our individual thoughts, words, and deeds.

The Functions of the Neuron

There are four basic roles for neurons in the nervous system. First, there are sensory neurons that carry information from the sense organs, such as the eyes and ears, to the brain.

Second are the motor neurons that control the voluntary muscle activity such as those used in speaking and moving arms and legs. They carry messages from the neurons in the brain to the muscles.

Third are the interneurons. Their function is to connect parts that are inside and outside the central nervous system. In essence, *interneurons are the central nodes of the neural circuits,* enabling communication between sensory or motor neurons and the central nervous system. They *play a vital role in reflexes* by quickly transferring information up and down the line.

Finally, are the specialized neurons in the brain. There are *hundreds of different types.* In general, their function is to receive, and process internal and external input received by the body and ensure that the body continues to function properly.

Individual neurons do not perform these functions on their own, but *the collective of neurons, working together,* do. *Researchers are still engaged in developing a scheme to neatly classify all these different neurons, but they are not yet close to a solution.*

How Neurons Communicate

Crucial to the functions of the brain and the nervous system is *the communication between the neurons* in the organism. Neural transmissions occur when a neuron is activated and sends out an electrical impulse along its axon, as shown in the Figure below. The upper neuron sends an impulse, or electrical spike, along its axon that connects the axon's terminals to several dendrites of the other neuron near the bottom of the Figure.

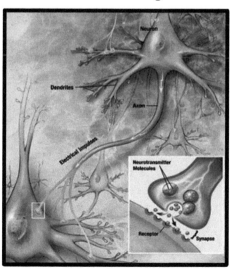

Neural Communication System

An axon terminal makes the connection to a dendrite through a *synapse* as shown in the *inset diagram of the Figure above. A synapse is the junction between the axon and the dendrite.*

The *synaptic cleft* is part of the synapse. *It is the gap across which signals flow from the axon of the transmitting neuron to the dendrite of the receiving neuron.*

How the signal crosses the gap can be *a chemical or electrical process. The chemical process is the most common type.* When the signal of energy arrives at the axon terminal, it provokes a synaptic sac containing *neurotransmitter molecules* to release those molecules.

Neurotransmitters are like chemical words, sending "messages" from one neuron to another. There are many different sorts of neurotransmitters: some stimulate neurons, making them more active; others inhibit them, making them less active.

The neurotransmitter molecules then cross the synaptic cleft and bind to chemical receptors located in the dendritic membrane of the other neuron. *This creates an impulse in that dendrite and thus continues the signal on into the other neuron.*

Electrical synapses are common in the mammalian brain. They differ from chemical synapses in several ways. In electrical synapses, the nerve impulses are transmitted by means of an *ion flow. This is a much faster process than found in the chemical synapse.*

The gap between the neurons is much smaller in an electrical synapse. About one tenth the size of the gap in a chemical synapse.

As shown in the below diagram, gap junction channels connect the two neurons in the *much smaller electrical synapse cleft.*

The electricity we use in our homes is generated by the motion of free electrons that exist in a copper wire. But the electricity generated by neurons results from the motion of the free positive ions[VIII] of sodium and potassium as they move across the cell membrane. The ions flow through the gap junctions which provide a low-resistance pathway for the ions. The ion flow is almost instantaneous, resulting in a transfer of the impulse in less than 0.3 microsecond.

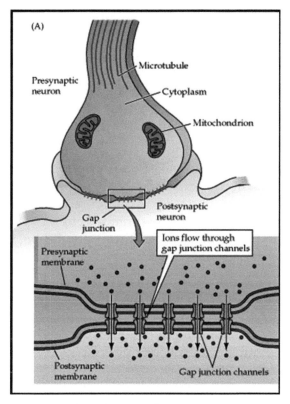

An electrical synapse

Synaptic Pruning

Synaptic pruning is a natural process that occurs in the brain between early childhood and adulthood. During synaptic pruning, the brain eliminates extra synapses. It is the brain's way of

---

[VIII] **Positive Ion** – Occurs when an atom loses an electron. The electron has a negative charge. So, then the atom has more protons (positive charges) than electrons, and the atom, therefore, has *a net positive charge.*

removing connections that are no longer needed. This makes the various brain functions more efficient as we get older.[57]

The reason the brain needs to prune synapses at this point in a person's development is that, during infancy, the brain experiences a large amount of growth. The growth is necessary to support the child's enhanced learning and memory function. As a result, there is an explosion of synapse formation between neurons called synaptogenesis.

Then, at about two to three years of age, the number of synapses hits a peak level, and the brain starts to remove synapses that it no longer needs. Once the brain forms a synapse, it can either be strengthened or weakened. Synapses that are more active are strengthened, and synapses that are less active are weakened and ultimately pruned.

Early synaptic pruning is mostly influenced by our genes. Later on, it is based on our experiences. So, whether or not a synapse is pruned is influenced by the experiences a developing child has. Constant stimulation causes synapses to grow and become permanent. But if a child receives little stimulation, the brain will keep fewer of those connections.

Synaptic pruning continues through adolescence, but not as fast as in earlier years. The total number of synapses begins to stabilize at this time. Then, a second pruning period occurs during late adolescence. According to the latest research, it appears that synaptic pruning actually continues into early adulthood and stops sometime in the late twenties.

Interestingly, during this time the pruning mostly occurs in the brain's prefrontal cortex, which is the part of the brain heavily involved in the decision-making processes, personality development, and critical thinking.

If the brain does too much pruning, or not enough pruning, some serious medical issues can develop. Research has shown a link between schizophrenia and "over pruned" brains.[58] Similarly, a link has been shown between "under pruned" brains and autism. More research is required to prove these connections, but if found to be true it could ultimately lead to new treatments for these problems.

Role of Neurons in Intelligence

For many decades, neuroscientists have been studying the relationship between the size, structure and functioning of the human brain and the measured level of intelligence for the individual.

Up until recently, however, it was not possible to *explore whether the fine structure[IX] of neurons is correlated with variations in human intellect.* The problem was in obtaining the necessary samples of living brain tissue to study.

But a few years ago, researchers at a Dutch university had the opportunity to examine *samples of live, temporal cortex tissue removed during surgeries* of cancer and epilepsy patients.

The lead researcher was Natalia Goriounova of the Vrije Universiteit Amsterdam. *She and her colleagues report that the microscopic anatomy of neurons and their physiological characteristics are **indeed linked** to individual differences in IQ scores.*[59]

The first step in their research process was to take a pre-operative, MRI scan of each patient to *measure the thickness of the temporal lobe cortex.* The researchers found that *higher IQ scores correlate with a thicker temporal cortex.*

Then, for each patient, two or three *pyramidal neurons* from the upper cortical layers were selected. Pyramidal neurons, so named for the pyramidal, or triangular, shape of their cell bodies, account for about *eighty percent of the neurons in the cortex. They are relatively large neurons.*

When researchers measured and compared the neurons harvested from the patients, they found that *larger, longer, more branched dendrites explained approximately twenty-five percent of the variance in IQ scores between individuals in a sample of twenty-five patients.*

The reason these characteristics make a difference in intelligence is that multiple, longer dendrites have extra surface area, which

---

[IX] **Fine structure** - the composition of a neuron as viewed on a small scale and in considerable detail.

could help increase the *number of synapses the neuron can form*. With more of these connections, the pyramidal neurons can produce an *output signal* that *integrates more inputs to neighboring neurons in a given time*.

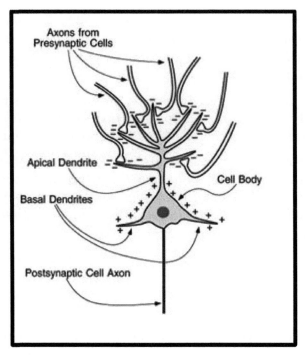

Pyramidal Neuron from Cerebral Cortex

The researchers also developed computational models that were used to explore how changes in the morphology (shape and density) of dendrites might influence the way the neurons worked. The analyses showed that *pyramidal neurons with larger, more dense dendritic trees fire more quickly, which allows them to transmit information faster.*

Given pyramidal cells that are larger, with more elaborate dendritic networks, the cortex may be significantly thicker. In turn, this would lead to more rapid information processing, and ultimately, increased intellectual performance.

The investigators also *recorded the activity of cells* in the brain slices they obtained from thirty-one patients. *These recordings showed that higher IQ scores were associated with neurons firing a little more quickly, especially during sustained neuronal activity.* Even this slight increase in how fast neurons pass on information can improve reaction times, and ultimately influence behavioral responses.[60]

The Centers of Intelligence in the Brain

Introduction

For many decades, researchers have sought ways to relate a patient's biological traits to their level of intelligence. For example, many scientists tried measuring head size and other phrenological characteristics, but of course, none of these were found to be useful.

Since the 1970s, however, the availability of advanced, neuroimaging technologies *have allowed researchers to study the brain during active cognitive functioning with much more success than in the past.*

For example, *Positron Emission Tomography (PET)* uses trace amounts of short-lived, radioactive material to map functional processes in the brain. When the material undergoes radioactive decay, a positron[X] is emitted, which can be picked up by the detector.

So, as the subject works to solve a problem, the areas of the brain that are involved in the cognitive process are highly radioactive, emitting more positrons. And those areas are noted.

Another useful technology for measuring brain activity involved in intellectual processing is *functional magnetic resonance imaging (fMRI).* It is a technique that *works by detecting the changes in blood oxygenation and flow that occur in response to neural activity.* This change occurs because, when a brain area is more active it consumes more oxygen, and, to meet this increased demand, blood flow increases to that area. This response shows up on the fMRI screen and can be used to produce activation maps which identify which parts of the brain are involved in a particular mental process.

[X] **Positron** - a subatomic particle with the same mass as an electron and a numerically **equal, but positive charge**.

The P-FIT Model of Intelligence

In 2007, Rex Jung of the University of New Mexico and Richard Haier of the University of California (Irvine) did a comprehensive review of the results of *thirty-seven neuroimaging investigations related to intelligence.* These studies, which had been conducted with a *total of 1,557 participants.* They included several measures of intelligence *including fluid and crystallized intelligence, reasoning, g, and games of reason including chess and GO.*

These inquiries identified *several, discrete brain regions* that were involved in intelligence, and how they were integrated during processing.[61] Jung's and Haier's findings suggest that intelligence is related, not so much to brain size, *but to how efficiently information travels through the brain.*

In turn, this is related to the number of neurons and their interconnections through axons, dendrites, and their associated synapses.

In their review, Jung and Haier compiled a list of all the brain areas that the thirty-seven studies had found to be related to intelligence. They placed greater emphasis on those areas that appeared multiple times.

*They found that most of the **brain areas thought to play a role in intelligence are clustered in the frontal and parietal lobes.***

The figure below shows the specific areas of the frontal and parietal lobes where the authors found information processing related to intelligence. The numbered areas of the brain are *Brodmann's areas,* as shown in the second figure below.

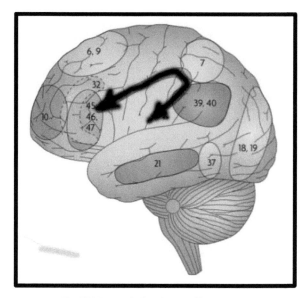

P-FIT model of Intelligence

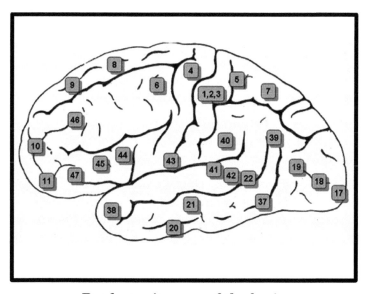

Brodmann's areas of the brain

Korbinian Brodmann was a German neurologist who became famous in 1909 for mapping the cerebral cortex and defining *fifty-two distinct regions, known as the Brodmann areas.* He identified the different regions *based on cytoarchitecture.* This is study of the cellular composition of the brain tissue under a microscope. It reveals *the characteristics of the cells* and other variations in the structure of the brain.

The data from the neuroimaging studies led Jung and Haier to *conclude that some of the brain areas related to intelligence are the same areas involved with attention and memory, as well as more complex functions like language.*

The authors believe that this integration of cognitive functions *suggests that intelligence levels might be based on* **how efficiently the frontal-parietal networks process information.**[62]

Based on these findings, the authors proposed a theory to match the evidence. *They called it the* **Parieto-Frontal Integration Theory (P-FIT) of intelligence.**

More specifically, they proposed that **general intelligence uses particular brain regions and incorporates**:

1. *Sensory processing,* primarily in the visual and auditory areas of the brain, including specific temporal and parietal areas.

2. *Sensory abstracting and elaboration* by the parietal cortex.

3. Interaction between the parietal cortex and frontal lobes for hypothesis testing of available solutions.

4. Response selection and inhibition of competing responses by the *anterior cingulate, found in Brodmann areas 24, 32, and 33.*

This P-FIT theory proposes that *greater general intelligence in individuals results from the* **greater communication efficiency** *between the particular areas identified, including the dorsolateral, prefrontal cortex, parietal lobe, anterior cingulate cortex, and specific regions of the temporal and parietal cortex.*

A 2010 review of the neuroscience of intelligence in *Nature Reviews: Neuroscience* by Deary, Penke and Johnson described *P-FIT as "the best available answer to the question 'where in the brain does intelligence reside?'".*[63]

Collective Human Intelligence

When researchers look at the *relationship between intelligence and learning in humans,* it is clear that high IQ individuals are also good at learning and acquiring knowledge. *There is a strong correlation between the two concepts,* which is intuitively obvious.

We know that we learn academic subjects by studying information from reliable sources like textbooks and scholarly papers. But an everyday means to *find out useful information quickly is through observing and listening to others.* It is called *social learning* and is a collective means to acquire valuable skills and knowledge.

Over the last twenty years, the internet and social media have magnified this process beyond measure. Virtually anything you want to learn about can be found on the internet. You can almost always find a YouTube video uploaded by some expert that shows you how to solve a particular problem like setting up a new piece of electronic hardware. Or the manual you need for that old tape recorder.

So *social learning is seen as a source of collective intelligence.* It can be defined as making smart decisions with groups of individuals contributing to the conclusion.

For example, it is common to find websites frequented by enthusiasts of a particular car, like a Nissan 350Z roadster. When one member has a problem in diagnosing an issue with his car, he goes online and poses the question to the other members. He quickly has several useful responses that usually allows him to solve the problem.

This ability is especially useful for making choices in everyday life like which restaurant to select for dinner, or the best pair of running shoes to purchase, or the candidates I should vote for in the next election.

But social learning can also go wrong. It is possible to also have collective "madness". This occurs when *incorrect, destructive knowledge* goes viral and is picked up by many members of the group through the social media. Scientists call this *maladaptive herding.*

"Herding is a form of convergent social behaviour that can be broadly defined as the alignment of the thoughts or behaviors of individuals in a group (herd) through local interaction and without centralized coordination."[64]

It is this type of behavior that sometimes triggers stock market instability. Or causes many people to adopt with religious fervor a wrong, destructive position on some subject, like "Climate Change".

So, instead of studying the facts and making their own decision, they take a firm position based on the flawed information they heard from others. And then pursue unswerving actions based on that poor decision.

To better understand *how collective intelligence can go wrong*, a group of researchers at the University of St. Andrews in the U.K. conducted a study.[65] *The objective was to determine if humans with a system to share social information could be flexible in dealing with it and avoid maladaptive herding.*

The researchers focused on two, related human behaviors in the experiment:

1. Conformity – that is, the extent to which an individual follows the majority's opinion.

2. Copying – the extent to which an individual ignores their own personal knowledge and relies solely on following others.

The experiment examined the patterns of human social learning through an interactive, online experiment with 699 participants, varying both task uncertainty and group size. In the online game, the objective for the player was to identify the best "slot machine" of those available, and to do it as quickly as possible, so as to win the most amount of money.

The experiment was designed so that players could see in real time what other participants were doing in the game. Then they could copy or ignore the choices of the others.

One conclusion from the study was that *a demanding task elicited greater conformity* in the responses of the group of participants. This suggests that when large groups are confronted with tough challenges, *collective decision-making becomes inflexible, and maladaptive herding behaviour is prominent.*

The popular, slot machine choice in the game got more popular because people followed the majority choice, even if it was not actually the winning machine.

The researchers also found that when the group size was small, or there was a less challenging version of the task, *the players were more flexible and were able to switch to a new and better strategy.* Thanks to this lower conformity in the smaller group, more people were willing to explore less popular options. As a result, they eventually found the best one as opposed to the one most chosen.

The major conclusion of the study concerning collective intelligence was that there is a high risk of maladaptive herding in large groups dealing with a difficult problem. People need to be aware of the risk and take minority opinions into account.

Stimulating independent thought in individuals is another way to reduce the risk of collective madness. *Dividing a group into subgroups or breaking down a task into smaller, easier steps helps to promote flexible, yet smart, human "swarm" intelligence.*

Consciousness

The Mystery of Consciousness

Introduction

We all know what consciousness is. We wake up in the morning to the sound of the alarm clock and are, once again, fully aware of ourselves. We still feel groggy, hate the sound of the alarm, and push the snooze button. And while we are lying there, we are once again mindful of the outside world; the peaceful sight of the sun streaming in through the window, the sound of the birds singing in the trees outside, the dog licking our face. We sort of remember the dreams we had, and how they felt like very real experiences. All these feelings are the "qualia" that represent our awareness.

Being conscious is obviously fundamental to intelligence. In fact, there is a theoretical basis that supports the idea that intelligence and subjective experience are related at a basic level.

Most neuroscientists agree that intelligence and consciousness are both functions of the brain. They know this from a large body of correlations between the brain's activity, and our experiences with intelligence and consciousness.

For example, people with traumatic injuries to the brain may have significant differences in their measured, *post trauma* intelligence or levels of consciousness. The scientists conclude *therefore, that brain activity creates intelligence and conscious experience.*

Unlike the proven neurological mechanisms of intelligence in the brain, as described by the P-FIT model, *it is still very much of a mystery to most neuroscientists exactly how the brain produces the conscious experiences that we have.*

In 2004, eight neuroscientists believed it was not then possible to describe how "consciousness" occurs, based on neuroscientific principles. They wrote an apology in the book, **Human Brain Function**:[66]

> *"We have no idea how consciousness emerges from the physical activity of the brain, and we do not know whether consciousness can emerge from non-biological systems,*

such as computers... At this point the reader will expect to find a careful and precise definition of consciousness. You will be disappointed."

As a result, consciousness is the subject of much speculation. Many theories have been developed, (see below, *The Many Theories of Consciousness*) but as shown, there is still little consensus among the experts. *What consciousness is, remains a mystery.*

When we are asleep, we still have some level of consciousness. We can still feel pain, we can dream and, later, when we are awake, we know that time has gone by.

But if, like me, you have had one or more of the following experiences: gone under general anesthesia, been knocked unconscious in an accident, had a sudden cardiac arrest, you went down to a much lower level of consciousness than normal sleep.

In this imposed state, we do not dream, feel pain, or a have a sense that time has passed. When we finally come back to full consciousness, we know that the experience was much different than just waking up in the morning.

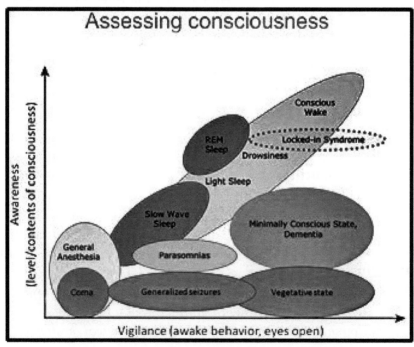

Above is a chart which shows the various levels of consciousness. The *horizontal axis* displays the *state of vigilance* of a person, the degree to which he is showing awake behavior. The *vertical axis shows his level of awareness.*

Consider the *Northeast corner* of the chart. That is the point with a high degree of Vigilance and a high degree of Awareness. It is, in fact, *our normal state of conscious wakefulness.*

From that point, if we go down toward the Southeast, just below normal wakefulness, is the very unfortunate condition of *"Locked-in syndrome"*. This is a rare, neurological disorder characterized by a high level of awareness and vigilance, but complete paralysis of voluntary muscles, except for those that control the eyes.

Continuing in the Southeast direction is sleep in its various stages, from drowsiness to light sleep to slow wave deep sleep. *Rapid Eye Movement* (REM) sleep is the level where we have most of the dreams that we remember. It can last for ten minutes to one hour. *During REM sleep, the brain is much more active than during other stages of sleep.*

In the *lower right corner* of the chart is a person who is experiencing *dementia or is in a vegetative state.* They have , an extremely low level of awareness but high vigilance.

Parasomnias is a sleep disorder characterized by abnormal behaviors such as *sleepwalking or night terrors.* It is low on the awareness scale, but medium on the vigilance scale.

Finally, in the *lower left-hand corner*, is extremely low awareness with extremely low vigilance. This characterizes a patient who is under general anesthesia or in a coma. I had multiple, personal experiences with this stage of the phenomenon.

I was in a recovery room at a hospital after having had a routine, heart diagnostic procedure. The last thing I remember was that I was sitting up in the bed, checking my cellphone. The nurse who was monitoring me was close by at her desk doing some paperwork. Then everything went black. I had a sudden cardiac arrest.

This means that my heart had abruptly and unexpectedly stopped beating. As a result, my blood stopped flowing, I stopped breathing, my pupils dilated, and I lost consciousness within seconds. I had flatlined and, for all intents and purposes, I was dead.

Currently, a patient who suffers a sudden cardiac arrest *in a hospital has about a   twenty-two percent chance of being resuscitated* and surviving to discharge. That means that four out of five patients who experience such an event do not make it. Pretty bad odds!

And it is *much worse outside a hospital.* With no expert medical personnel, or an Automatic External Defibrillator close at hand, *less than 10% of sudden cardiac arrest patients survive.*

You may remember a movie made in 1990, re-made in 2017, called "Flatliners". In that film, five medical students experiment by using drugs to purposely have a sudden cardiac arrest. Their idea was to have a "near death" experience and get a glimpse of the afterlife before their friends revived them.

But in real-life, not all patients who survive their cardiac arrest have a near-death experience. It is *only about one in five of them that do.*

Over the years, there have been hundreds, if not thousands, of anecdotal reports of near-death experiences from many countries and cultures. The similarities in these stories raised the interest of the scientific community, and, as a result, *there have been several, rigorous, scientific studies of the phenomenon.*

For example, the renowned, international medical journal, **Lancet**, published a study of near-death experiences by patients with sudden cardiac arrest. It was conducted by a Dutch cardiologist, Pim van Lommel.[67]

The experiences of his patients tended to be like many of the other recorded, near-death experiences found in the literature. They often entail *being conscious, but knowing you are dead, and finding yourself in a dark tunnel with a bright light at the other end. When the subject emerges into the light, they find themselves in a most beautiful place, and having feelings of total serenity, security, warmth, and unconditional love.*

Many experiencers also see long-dead relatives, friends and even pets. The patient can interact with them, and it all seems very real and comforting. *Many of these patients claim afterwards that their experience changed their lives in a tangible and positive way.*

Other sudden cardiac arrest patients claim to have a different kind of incident. It is an *out-of-body experience that also feels very real to them.* In these cases, the patient believes that they are floating above their hospital bed, looking down on the scene where the medical staff is trying to resuscitate them. Many patients can recount the actual actions and conversations that took place, confirmed by the participants. Yet, they had flatlined, their brains were completely inactive, and they had no senses.

Many doctors or scientists have tried to develop a reasonable, medical explanation for these experiences. All sorts of theories have been proposed, but there is no scientific consensus on how these events can happen.

One of the most interesting aspects of near-death experiences is the idea of "consciousness" and how it relates to this mysterious phenomenon. Most dictionaries define consciousness as "the fact of awareness by the mind, of itself and the world." But when blood stops flowing to the brain because of cardiac arrest, the victim's brain activity ceases and they quickly lose "consciousness". *Yet, these victims still claim to have conscious experiences.*

Van Lommel concluded in his Lancet article that the results of his study demonstrate that the patient's awareness did indeed continue after all brain activity stopped. In his summary, he raised the question, "If consciousness is a function of brain activity, as most scientists believe, how could consciousness be experienced after the brain no longer functions?"

von Lommel answered his own question in a later book that he wrote. It was titled **Consciousness Beyond Life**[68]. He claimed in the book, like some philosophers have in the past, and a few today, that *human consciousness functions independently of the brain.*

The brain, he says, *"is just the receiver of the signals generated by this higher-order consciousness, much as a television set is the receiver of signals from a television station."* This concept, some philosophers say, is evidence of the existence of the soul, and the near-death experience provides a glimpse of the afterlife. In the

Flatliners movie, it is this glimpse that the student doctors were seeking.

Other doctors, trying to explain the conscious experiences of cardiac arrest patients, attribute it to the act of heart massage in the resuscitation process. The result of the massage, they say, is that there is some blood flow to the brain. And that is what explains the consciousness required for the near-death experiences.

On the other hand, von Lommel's view is supported by many neurologists. They say that until the heart starts beating again, there can be no brain function.

Few neuroscientists support the idea of consciousness operating independently of the brain. One that does is Dr. Donald Hoffman, Professor of Cognitive Science at the University of California, Irvine.

Hoffman is working on developing a mathematical, scientific analysis to prove his theory. He calls it *Conscious Realism.* It is discussed later in this Part as one of "The Many Theories of Consciousness."

I am obviously among the one in five patients who suffers a sudden cardiac arrest in a hospital and survives. After everything went black, the next thing I knew, I was looking up at a bearded nurse who I later learned was named Eddie. Eddie, and the other medical professionals in that recovery area, went into quick action and brought me back within a few minutes. As I returned to fuller consciousness, I remember the crowd around me and the frantic looks on their faces, but nothing more. To me, it felt like having descended into blackness, and then nothingness. There was, indeed, no feeling of time having passed. I experienced nothing of the "afterlife" as did some other near-death survivors.

This was very disappointing. Like the doctors in "Flatliners", I would have enjoyed having such an experience. Since it is only about 10% of sudden cardiac arrest patients who report near-death experiences, some doctors have theorized that everyone has the experience, but, like our REM state dreams, the majority of people do not remember.

The next day, I had an open heart, surgical procedure using general anesthesia. I went "under" at nine am in the morning, and the next thing I knew, it was one am the next day, sixteen hours later. But it could have been one am, five years in the future and I would not have known the difference.

The "Hard Problem" of Consciousness

Before reading this book, you may not have thought much about consciousness. You might figure that it is just your mind at work. And you presume that your mind originates in your brain and, of course, that scientists understand all about it.

But that is not the case. Unlike the research on the locations in the brain associated with intelligence, there is still a great debate going on among researchers and philosophers about *exactly what consciousness is and how it happens.*

One side of the argument is called *philosophical materialism.* With this view, the mind and consciousness are purely the result of the biochemistry of the brain and nervous system.

A different idea is *Dualism.* This is the belief that the body and consciousness are separate. As discussed above, this is the view of Dutch cardiologist, Pim Van Lommel.

David Chalmers is a philosopher and cognitive scientist. He is also a Professor of Philosophy and Neural Science at New York University. It was he who introduced the term "hard problem of consciousness".[69]

Professor David Chalmers

He contrasts understanding consciousness with the "easy problems" of neuroscience. These are the ones involving the eventual, neuroscientific understanding of the many other functions of the brain, like focusing attention on a particular activity, multi-tasking, planning, recognizing language, and forming memories.

Another example of an "easy problem" was locating the areas of the brain involved in intelligence, and understanding the mechanisms for its functioning, as in the P-FIT model.

On the other hand, the "hard problem", according to Chalmers, and agreed to by most neuroscientists, is trying to explain why and how we have *the conscious experiences that we do.* For example, what we see and feel about an exquisite red rose, with its delicate petals and sweet aroma, is a subjective experience, not an objective one.

Philosophers call these *qualia,* or individual instances of subjective, conscious experiences. We know what we think of as red or blue, and how we feel about those colors, but we cannot know if other people experience them in the same way.

To be clearer, the "hard problem" of consciousness refers to the question of *how, exactly, does the physical brain give rise to our unique experiences.* Certainly, everyone agrees that the brain takes in, processes, and interprets concrete information from the outside world. But individuals can *feel very differently about the same information.*

For example, two fans are watching a football game. They see and hear the same play as it occurs. The visual and auditory centers of their brains take in the information, and each fan knows exactly what happened. But if the two fans are rooting for opposing teams, *they will feel very differently* about the results of that play.

Because of these different feelings, despite the same inputs, some experts contend that we will never be able to "locate" consciousness entirely within the structures and chemical processes of the brain. For those scientists, consciousness is too complex to explain completely in terms of gray matter and brain chemicals.

The brain, they conclude is necessary for consciousness, but is not sufficient. As you will note in the discussions below, not all scientists agree with them.

There are several credible researchers who have considered the evidence and concluded that there is a way to explain how consciousness occurs. In *The Many Theories of Consciousness*, below, their arguments are presented. Yet, there is still no theory that stands out as being universally accepted.

Quantum Mechanics and Consciousness

What physicists do for a living is to observe what is going on in nature, and then develop predictive, mathematical descriptions of those actions. They can then use the mathematics to show how those activities will change over time and as conditions change.

Classical Mechanics is an example. It is a branch of physics developed by Isaac Newton and others starting way back in 1687.[70] It is a physical theory that describes the motion of macroscopic objects, like, for example, the motion of a ball hit by a bat.

Newton's mathematical equations make it possible to determine how the ball will move once it leaves the bat. These equations work very well and make excellent predictions about the position of the ball at any point in time. *The same holds true for an iron ball being fired from a cannon.*

However, when physicists started working with very small particles, like atoms and electrons, they found that Classical Mechanics was not able to describe the observed behavior of the particles. *They needed to have a new theory.*

What they noticed in their experiments was that an electron, a small, negatively charged particle, in some cases behaved like a particle as expected, but *in other situations it behaved like a wave!*

But, beyond this observation, the most startling thing about Quantum Mechanics[XI] to physicists, when they first started to study subatomic particles, **was that consciousness played a role in the results of their experiments**

The Classic Double Slit Experiment

A good example of what happens at the quantum level, and caused the physicists to revise their thinking, is *found in the classic*

[XI] **Quantum mechanics** is a fundamental theory in physics that provides a description of the physical properties of nature at the scale of atoms and subatomic particles.

"double slit experiment". Two experimental arrangements are shown in the diagram below.

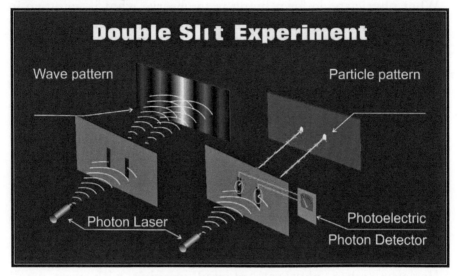

The Double Slit Experiment

In this first test, illustrated on the left of the Figure, a *coherent* light source, such as a laser beam, illuminates a plate that is pierced by two parallel slits. The light wave passes through the slits.

This creates two new waves that emerge from the other side of the slits as shown. The waves travel away from the slits on the other side of the plate and *begin to merge with each other* as shown in the diagram.[XII]

At the points where the merging waves meet and *their crests line up* together, there is *constructive interference.* The *resulting wave has a higher amplitude,* or more intensity.

Where the *crest* of one wave meets the *trough* of another, the result is *destructive interference,* and the *resulting wave has a lower total amplitude close to zero, or no intensity.*

[XII] If electrons are used in the experiment instead of light, the *same type of wave interference pattern* is observed. The electrons are, therefore, behaving as a wave, not a particle. **But**, when a particle detector is in place, and the electrons are observed before they reach the slits, *they behave as particles and go straight through!*

The result of this first double slit experiment is observed on a screen on the left side of the diagram. In the Figure above, *the pattern that appears on the screen* is alternating light and dark bands, with the brightest band being in the center. The light bands are the result of constructive interference, while the dark bands in between are from destructive interference.

*Now, in a **second experiment, illustrated on the right** of the Figure, all is the same **except for a photon**[XIII] **detector being placed near the slits** in front of the plate. It is prepared to watch what passes through the slits.*

When the laser is activated, *surprisingly **the detector sees photons**, or light particles, **instead of waves passing through the slits.** **And each photon** passes through only **one** slit (as would a classical particle), and **not** through **both** slits (as would a wave).*

*The pattern seen on the right screen is very different. Instead of a series of light and dark bands generated from waves interfering with each other, the pattern becomes **just two bright bars** where the photons went through the slits and directly to the screen! Exactly as you would expect if light were made up of "particles" instead of waves.*

*Physicists now believe that, because the light wave was **observed by a conscious human** who is monitoring the photon detector, its **wave function "collapsed",** and it appeared and behaved as a particle called a photon.* **This may be the strangest and most profound finding in all of quantum physics.**

The results of this experiment deeply troubled the physicists. It seemed to undermine the basic assumption behind all science: that there is an objective world out there, irrespective of us. If the way the world behaves depends on how – or if – we look at it, what can "reality" really mean?

[XIII] A **Photon** is a particle representing a *quantum*, or discrete bundle, of light or other electromagnetic radiation. A photon carries energy proportional to the radiation frequency *but has zero rest mass*.

More about this idea later in this Part when we discuss the work of physicist John Archibald Wheeler. As well as *the section on the theory of* **Conscious Realism** *proposed by Professor Donald Hoffman.*

To summarize, a basic principle of quantum mechanics, when working with light or subatomic particles like *electrons*, is that these objects are *normally described by a "wave function".* How that wave function evolves over time is *described by Schrödinger's wave equation*[XIV].

Now, on the other hand, when the light, or the electron, is being observed by a conscious being, the wave function appears to "collapse" and the single particle is observed. **This phenomenon is called wave-particle duality** *and is a key concept of Quantum Mechanics.*

The concept of "wave function collapse" is called *the Copenhagen Interpretation of the double slit experiment.* It is because this idea was proposed by Neils Bohr and his associates who were based in Copenhagen, Denmark. Other physicists have suggested other explanations, but this is the one that is generally used by physicists in their calculations because it works.

Note that the scientists did not, and do not, understand how or why the wave function collapses. It just appears that way, and their calculations and predictions are correct, so they live with it.

[XIV] **Schrodinger's wave equation** - Erwin Schrodinger developed the equation. It is a linear, partial differential equation that describes the wave function, or state function, of a quantum-mechanical system. It is a key result in quantum mechanics, and its discovery was a significant landmark in the development of the subject.

The Many Theories of Consciousness

Orchestrated Objective Reduction Theory

The most accepted version of quantum mechanics is the *Copenhagen Interpretation.* This is the view that says that an electron is a wave until it is detected by a conscious observer. Then the wave function collapses, and it behaves like a particle. It was extremely hard, but *physicists had to accept the fact that consciousness plays a role in their theory.*

This fact, unfortunately, later led to mystics, psychics, and clairvoyants adopting the "quantum" idea for their beliefs. The term, "Quantum consciousness", became tarnished by New Age religions which attributed such things as telepathy and telekinesis to quantum mechanics. The result was that physicists became embarrassed by putting "quantum" and "consciousness" in the same sentence. In fact, it became career suicide for physicists to study anything to do with "consciousness".

But this did not deter a world-renowned, Nobel Prize winning physicist from Oxford University, Sir Roger Penrose, who is still active at the age of 90. His career was already well established in cosmology and general relativity, so he proceeded fearlessly.

Penrose believes that consciousness itself, in fact, has its origin in quantum physics. In his 1989 book, "The Emperor's New Mind", Penrose argues that the brain is far more sophisticated than a conventional computer, perhaps closer to a quantum computer. Instead of operating on a strictly binary system of on and off, the human brain works with computations that are in a superposition of different quantum states at the same time.

Together with Doctor Stuart Hameroff, an anesthesiologist, Penrose developed the *Orchestrated Objective Reduction theory (Orch OR)* of consciousness. Penrose and Hameroff propose that *consciousness originates at the quantum level inside the neurons.*[71] Orch OR posits that consciousness is based on *quantum processing performed by qubits.*

These qubits are formed collectively on minuscule protein structures in the neurons called *microtubules.*[XV]

Information and memory are stored, and quantum processed, in the microtubules. Penrose and Hameroff believe that *it is these quantum devices that are orchestrating our conscious awareness.*

Penrose and Hameroff

But recall that, to be stable, *quantum systems need to be coherent*, meaning that the system in superposition must be in related states. Therefore, for Orch OR to work, *coherence must be maintained in the environment of the brain.*

Because of this need for coherence, The Penrose-Hameroff theory is not appreciated by many other scientists. Critics, such as Max Tegmark, a physicist from MIT, in a study published in 2000[72], typified the major complaint of the detractors.

He said that the brain is too "warm, wet, and noisy" and cannot sustain a quantum process. The reason is *the decoherence* of the quantum systems that would occur in such an environment. In fact, decoherence is expected to be extremely rapid in warm and wet environments like living cells.

[XV] **Microtubules** are small tubes formed by thirteen filamentous strands. Each filament is composed of a chain of protein called tubulin. Microtubules in a neuron are used to transport substances to different parts of the cell.

Tegmark calculated that quantum superpositions of the molecules involved in neural signaling could not survive for even a fraction of the time needed for such a signal to get anywhere in the brain. *Thus, he ruled out the Orch OR theory.*

Steve Paulson, on the other hand, while writing for the online science publication, Nautilus, describes the 90-year-old Penrose's theory as an *"audacious—and quite possibly crackpot—theory* about the quantum origins of consciousness.

Penrose believes we must go beyond neuroscience and into the mysterious world of quantum mechanics to explain our rich mental life. Paulson says that no one quite knows what to make of this theory, but conventional wisdom goes something like this: Their theory is almost certainly wrong, but since Penrose is so brilliant ('One of the very few people I've met in my life who, without reservation, I call a genius,' physicist Lee Smolin has said), *we'd be foolish to dismiss their theory out of hand."*

Penrose and Hameroff continue over the years to respond to their critics and maintain the plausibility of their theory. But they still do not have many supporters.

The Attention Schema Theory

There are some scientists and philosophers who disagree with the idea of the "hard problem of consciousness" being unsolvable. They believe that it is possible to develop a scientific explanation for consciousness. One is Michael Graziano, a neuroscientist at Princeton University.

Professor Michael Graziano

121

For Doctor Graziano, trying to understand how the brain works at the level of neurons, synapses and axions is like trying to understand how a computer produces an answer to a problem by examining its transistors. He believes that to understand what is really happening, you must be looking at a much higher level of the brain's operations. *In other words, its "software programs".*

In Graziano's case, he believes that, to understand consciousness, you need to appreciate how the brain models its internal processes for controlling "attention". *He called it the "attention schema theory" (AST).*

In a nutshell, the theory proposes that subjective awareness, or consciousness, is just the brain's internal model of the process of attention.

Because there is an overwhelming amount of information coming constantly into the brain, it has developed, over millions of years of evolution, the capacity to focus its processing resources on some signals more than others at any point in time. The focus of attention may, at a particular moment, be on external, incoming sensory signals; or it may be on internal information, such as specific, recalled memories, or the emotions that you are feeling at that instant. *The ability of the brain to process select information in a focused manner is called attention.*

To use its attention capability effectively, the brain *builds a simplified, schematic model of the attention process which is continuously being updated.*[73] That representation, or *internal model*, is *the attention schema*. This top–down control of attention is improved when the brain has access to such a simplified model of attention itself.

This schema assists the brain as it monitors and directs its attention to its many functions. The template is like a cartoon sketch that depicts the most important, and useful, aspects of attention, without representing any of the mechanistic details that make attention actually happen.

The heart of the attention schema theory is that *there is an evolutionary, adaptive value for a brain to build the construct of awareness: it serves as this model of attention.*

It is the external and internal information in that model at any point in time that leads the brain to conclude that it has a non-physical essence of awareness.

And since we have access to our internal attention model, *we report the content of the model as though we are reporting literal reality.* We say that we are aware of our feelings because of the attention model.

Another important, evolutionary development in the AST was the idea that the personal attention schema model, or our self-model, *also became a model for us to understand how others might behave.*

What happened was that *our own attention schema was adapted to model the attentional states of others.* This provided valuable insights for social prediction. *Now, not only did the brain attribute consciousness to itself, but it also began to attribute consciousness to others.* This capability is what psychologists call *"the Theory of Mind".*

Data from Graziano's lab suggests that the cortical networks in the human brain that allow us to attribute consciousness to others overlap extensively with the networks that construct our own sense of consciousness. Graziano believes that the self-model and the social model evolved in tandem, each influencing the other.

We understand other people by projecting ourselves onto them. But we also understand ourselves by considering the way other people might see us.

These internal models do not provide a scientifically precise description of attention. In fact, the model is silent on the physical, neuroscientific mechanisms of attention. Instead, like all internal models in the brain, it is simplified and schematic for the sake of efficiency.

In a later chapter, we will be discussing the evolution of Artificial Intelligence (AI). Since most neuroscientists do not know what consciousness is or how we have it, they are not certain if superintelligent AI machines will develop consciousness on their own.

On the other hand, with the understanding of the Attention Schema Theory (AST), scientists will be able to program AST into the AI. And for all intents and purposes, the AI will report that it is self-aware and is conscious. And it will act accordingly.

The Integrated Information Theory

The "hard problem" of consciousness, according to David Chalmers, is trying to explain, using the principles of neuroscience, why and how we have *the conscious experiences that we do.* Many scientists wonder if they will ever solve this problem.

So, the creator of the Integrated Information Theory (IIT)[74], Giulio Tononi of the University of Wisconsin, takes a different approach. He *accepts the existence of consciousness* and reasons about *the properties that a hypothesized, physical neural substrate, the brain, would need to have to account for it.*

Giulio Tononi

IIT proposes that a single conscious experience is made up of a collection of 'concepts', which are created by specific arrangements of integrated information. For example, sensory information about an apple (the sweetness of its taste, the color red, the crunchy texture) **integrate to form the concept of an apple**.

More specifically, coordinated firing in various brain regions dedicated to taste for the sweetness of an apple, visual processing for the redness, and tactile sensation for the crunchy texture of the apple, *come together to form the conscious concept of an apple.*

IIT further states that this phenomenon occurs on a larger scale within the brain to form the consciousness experienced by the *individual during any particular moment of life. These concepts are integrated together to form a **maximally irreducible conceptual structure** (MICS), which ITT defines as consciousness.*

IIT has not been universally acclaimed by scientists studying consciousness. As a scientific theory of consciousness, they say that, by its own definitions, it cannot be proved to be true. It can only be determined to be "either false or unscientific".

It has also been *denounced by other members of the consciousness field as requiring "an unscientific leap of faith".* Furthermore, the idea has been derided for failing to answer the basic questions required of a theory of consciousness.

Philosopher Adam Pautz says, "As long as proponents of IIT do not address these questions, they have not put a clear theory on the table that can be evaluated as true or false."

Conscious Realism

There are a few neuroscientists exploring the *radical idea that consciousness itself is fundamental in the Universe.* One of these is Donald Hoffman, a full professor of cognitive science at the University of California, Irvine, where he studies consciousness, visual perception and evolutionary psychology using mathematical models and psychophysical experiments.

Donald Hoffman

Most physicists agree that *"unconscious" particles and fields that obey the laws of quantum physics are fundamental in the Universe.* On the other hand, *Hoffman suggests instead that it is "conscious agents" that are fundamental.* This is a radical departure from accepted science.

Hoffman proposes *"Conscious Realism"* as a philosophy and scientific theory.[75] As he describes it:

> *"Conscious realism asserts that the objective world, i.e., the world whose existence does not depend on the perceptions of a particular observer, consists entirely of **conscious agents**."*

Conscious agents, according to Hoffman, are *the primitive constituents of reality. The objective world, he says, consists of conscious agents and their experiences.*

Hoffman says in the Abstract to his paper describing Conscious Realism:

> *"Despite substantial efforts by many researchers, we still have no scientific theory of how brain activity can create, or be, conscious experience. This is troubling, since we have a large body of correlations between brain activity and consciousness, correlations normally assumed to entail that brain activity creates conscious experience.*
>
> *Here I explore a solution to the mind-body problem that starts with the **converse assumption**: these correlations arise because **consciousness creates brain activity, and indeed creates all objects and properties of the physical world.***
>
> *To this end, I develop two theses. The multimodal, user interface theory of perception states that perceptual experiences **do not** match or approximate properties of the objective world, but **instead** provide a **simplified**, species-specific, **user interface** to that world.*
>
> *Conscious realism states that the objective world consists of conscious agents and their experiences; these can be mathematically modeled and empirically explored in the normal scientific manner."*

In the paper, he expands on these ideas:

> *"So, what is space-time and what are our perceptions of objects? I think a good way to think about them is that they*

are just a user interface. (like the graphic user interface (GUI) used for computer applications) *We* (humans) *evolved a user interface (* through Darwin's theory of evolution). *If you are crafting an e-mail on your computer and the icon for the e-mail is blue and rectangular and on the right corner of your screen, that doesn't mean that the e-mail in your computer is blue and rectangular and in the right corner of your computer."*

Furthermore:

"The interface is not there to resemble reality. It is there to hide reality and to give you eye candy (visual images that are superficially attractive and entertaining but intellectually undemanding) *that lets you do what you need to do. That is what evolution did.*

3D space is our desktop. Physical objects are the eye candy. **They are there not to show us the truth but to hide the truth and let us act in ways that keep us alive."**

What Hoffman is saying is that, *in evolutionary terms, a species does not require to see and understand all the details of any given environment* — or it does not require "truth", as Hoffman says. **All that is needed is just enough information to survive and propagate.**

Expanding the ideas further, Hoffman goes on:

"According to conscious realism, when I see a table, I interact with a system, or systems, of conscious agents, and represent that interaction in my conscious experience as a table icon. Admittedly, the table gives me little insight into those conscious agents and their dynamics.

"The table is a dumbed-down icon, adapted to my needs as a member of a species in a particular niche, but not necessarily adapted to give me insight into the true nature of the objective world that triggers my construction of the table icon.

"When, however, I see you, I again interact with a conscious agent, or a system of conscious agents. And here my icons give deeper insight into the objective world: they

127

convey that I am, in fact, interacting with a conscious agent, namely you."

Most neuroscientists agree that Donald Hoffman's *thesis is a radical departure from the most accepted concept*, i.e., that somehow the brain creates our conscious experiences.

Yet, as Hoffman says, none of his colleagues have yet to produce a believable theory based on accepted neuroscience. So his approach is at least an attempt to find a workable, scientific concept that can be tested mathematically.

Wheeler's Participatory Anthropic Principle

John Archibald Wheeler was an eminent American, theoretical physicist and cosmologist with a wide range of interests. Based on his understanding of quantum physics, he became intrigued with *the role of consciousness in the physical world*. Ultimately, he developed *a theory of the universe* that included that idea. It is a highly controversial concept, but one *with support from some other physicists.*

Wheeler has an extraordinarily strong background and reputation in the field of physics. During the 1930s, as a young physicist, he worked with Neils Bohr, one of the founders of quantum mechanics, to explain the process of nuclear fission. Wheeler spent most of his productive years, however, *on the faculty of Princeton University.*

John Archibald Wheeler

One initial area of pursuit at this Ivy League institution was his *collaboration with Albert Einstein* in the old master's search for the unified theory of physics. This was Einstein's attempt to bring together his general theory of relativity with electromagnetism. *That quest, unfortunately, never bore fruit for either of them.*

Wheeler managed to spend the bulk of his Princeton career in the *field of cosmology.* This is a subset of astrophysics that *studies the origin and evolution of the universe.* He is best known for coining the terms *"black hole",* a region of space-time where gravity is so strong that not even light can escape from it; and *"wormhole",* which is a tunnel in space-time.

Wheeler was always fascinated by the ideas of quantum mechanics, including the role that consciousness plays in the physical world. And that intrigue carried over to his work in cosmology.

He was quoted as saying, *"Are life and mind irrelevant to the structure of the universe, or are they central to it?"* He concluded that it was the latter, **and that *observers are necessary to bring the universe into being. He called it the "Participatory Anthropic Principle" (PAP).***

Wheeler originated the notion of a *"participatory," conscious universe, a cosmos in which all of us are embedded as co-creators.* This concept replaced the accepted idea of the universe being "out there," and separate from us.

To prove his concept experimentally, he proposed *a variant of the double-slit experiment.* It was called *"the delayed choice"* test.

Wheeler wondered what would happen if the observer were to watch from the *back side* of the plate with the double slits *instead of in the front.* In other words, *if the intelligent observer were to watch the results of the event, would that have the effect of collapsing the wave function in the past before the photons went through the slits.*

You recall that, in the *original double slit experiment,* with no observer looking at the front of the double slit plate, light behaved as a wave and created an interference pattern of alternating dark and light bars on the screen, as shown on the left of the Figure below.

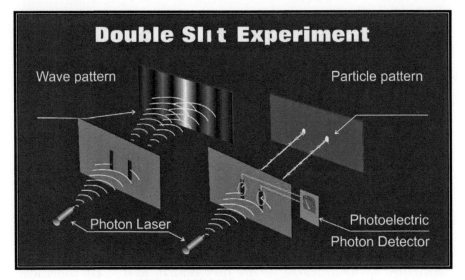

The Double Slit Experiment

However, when a photon detector was placed to monitor the front of the plate, the *light wave appeared to "collapse"* and then photons could be observed going through the two slits. Accordingly, a pattern of *double bright bands* appeared on the right screen rather than the interference pattern.

In Wheeler's proposal *to re-do the experiment, there was no observer in the front of the plate,* in the form of the photon detector, to collapse the wave. Instead, the detector was placed *on the other side of the plate* and would watch to see if the wave or particles *emerged from the slits.*

Wheeler speculated that the observer on the backside would *affect how the wave **behaved in the past**,* i.e., before it passed through the slits (a delayed choice). And, he believed, particles would be seen emerging from the slits. ***That is exactly what happened when his hypothesis was tried in a 2007 experiment.***

Wheeler said that, s*ince our observations in the present affects how a photon behaved in the past,* this **proves the concept that, human consciousness shapes not only the present, *but the past as well.***

Wheeler postulated, therefore, that an observers' consciousness was required to bring the universe into existence.

Many other physicists, but not all, agreed. They believe that *our bodies and brains become parts of the quantum mechanically described physical universe.* To John Barrow and Frank Tipler, in their widely controversial, 1985 book, *The Anthropic Cosmological Principle*[76,] this means that **the universe itself comes into being only if someone is there to observe it**. Essentially, the universe requires some form of life present for the wavefunction to collapse in the first place, meaning that *the universe itself could not exist without life in it.*

The Solms Theory: The Source of Consciousness

Yet another theory of consciousness has been proposed by Professor Mark Solms. He is the Director of Neuropsychology in the Neuroscience Institute at the University of Cape Town, South Africa.

His idea can be *described as a neuro-psycho-evolutionary vision of the emergence of mind.* While *it is a minority view among neuroscientists*, it has appeal as an explanation for how and why consciousness evolved.

He explains his theory in detail in his recent book, **The Hidden Spring**: *A Journey to the Source of Consciousness.*[77]

Mark Solms

It is commonly believed among neuroscientists that consciousness is a higher brain function, most likely centered in the cerebral cortex. Solms, on the other hand, postulates that consciousness is instead **rooted** *in the brainstem, which is among the "oldest" parts of the brain in the evolutionary sense.*

In particular, **he believes that the *source of consciousness* is centered in the *midbrain region of the brainstem (see diagram below).***

*His theory proposes that consciousness **is an evolutionary response to the human organism's need to minimize the energy it expends to meet its basic physical and psychological needs.*** Essentially, he is saying that *affect* consciousness leads to perceptual consciousness. *The affective phenomenal experiences*[XVI]*, or conscious feelings,*

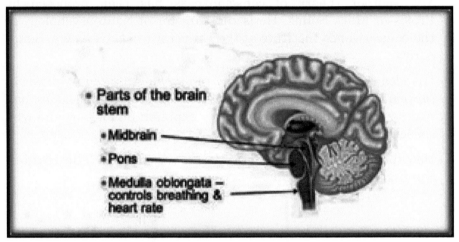

The Brain Stem Components

from the midbrain provide the "energy" for the developmental construction of higher forms of cognitive consciousness in the cerebral cortex.

Solms acknowledges that *perceptual* consciousness is a cortical phenomenon, although one that only exists when the cortex is suitably aroused by the RAS (reticular activating system), another brainstem region.

[XVI] *Affective experience*, in psychology, refers to the underlying experience of feeling, emotion or mood

In summary, Solms believes that the *biological function of consciousness is to **feel** our way through life's unpredicted problems using voluntary behavior.* The source of consciousness starts in the brain stem with our feelings about what is "good" or "bad" for us. It then links our internal affects (rooted in our needs) with the representations of the external world. And we take voluntary, conscious actions to meet our requirements.

Part II

Animal Intelligence

II. Animal Intelligence

The Power of Evolution

Introduction

Most of us feel positively about animals. You Tube videos, television programs and advertisements portraying all types of creatures are immensely popular. In fact, we feel so strongly about animals that the vast majority of us have a pet as part of our families.

According to the annual survey, conducted in 2020 by **The American Pet Products Association**, *sixty-seven percent* of U.S. households, or about eighty-five million families, own one or more pets. In the United States, the pets are primarily dogs and cats.

If you are a dog owner like me, you spend a lot of time with your "best friend". You get to know each other very well indeed. Certainly, communication is not an issue. The dog learns what you want, and you learn what he wants. With words and gestures on your part, or barks, whines, and poses on his part, we get the message across to each other. The canine is particularly good at interpreting our words and hand signals into a thought that makes sense to him.

One problem is that we tend to anthropomorphize our dogs. We see their behaviors as if they were human beings rather than a creature who likely thinks about the world in quite a different way.

Like humans, some dogs are smarter than others. We all believe our guy belongs at the head of the class. But, just how smart is he, compared to other dogs? How about compared to other species? That is a question that has also intrigued many scientists over the years.

For generations, therefore, *scientists have been studying comparative animal intelligence*. A key reason they have had in this pursuit is to *better understand the evolution of cognition* in all creatures including humans. In particular, how and why did the level of cognition in animals change over millions of years to reach our stage as humans?

Darwin's Theory of Evolution

"Evolution" means "change" in living organisms over long periods of time. Darwin understood that changes occurred, but he did not know about DNA and how it participated in the evolutionary process. Now we recognize that species evolve slowly because their DNA evolves slowly.

Understanding the basics of the "Theory of Evolution" is still an important starting point for learning about the evolution of animal cognition. So, that is covered below. In Part V, **The Future of Humankind**, a *detailed discussion of the evolution of man* is provided as a basis for that discussion.

In 1859, Charles Darwin formulated his "Theory of Evolution by Natural Selection". He published it in his landmark book, **On the Origin of Species**.[78]

In this work, he proposed a process that allowed living organisms to *change their heritable or behavioral traits over prolonged periods of time. These changes were such that the creatures became better adapted to their environments.* As a result, the organisms survived *and* had more offspring. *Thereby propagating the improved adaptations.*

Charles Darwin

The Theory of Evolution begins with the first life forms that appeared on Earth almost four billion years ago. This was only 500 million years after the Earth was formed by the accretion of space gas, dust, and other galactic materials.

The conditions on Earth at that time were extremely harsh compared to today, so *exactly how life first developed is still very much of a mystery to scientists.*

What is known is that the earliest life-forms were microorganisms that lived in the oceans. They were *single-celled, minuscule creatures found in hydrothermal vent precipitates.*

Hydrothermal vents are the result of seawater percolating down through fissures in the ocean crust. The *cold seawater is heated by hot magma and reemerges to form the vents.* Seawater in hydrothermal vents may reach temperatures of more than 700° Fahrenheit.

These hydrothermal vents were soon surrounded by significant numbers of organisms that thrived, even in that unforgiving environment. These biological communities depended upon the chemical processes that result from the interaction of seawater and the hot magma associated with underwater volcanoes.

The microbes reproduced and became more numerous. But they also competed with, and faced predation, by other entities. So, *the survival of individual organisms was threatened by the environment in which they lived.*

But a natural process occurred which changed everything. The microorganism's *DNA evolved as mutations occurred in some genes during the process of DNA replication.* These **mutations** resulted in the *creation of a different gene from the original.*

Some of these mutated genes, *by chance*, created a *new*, "beneficial" trait. Those creatures with this beneficial trait were more "fit" and better able to survive longer and, therefore, produce more progeny.

Those without the trait, were less fit and produced fewer progenies. Therefore, those creatures with the beneficial trait eventually grew dominant, while their less fit counterparts eventually died out.

Over long periods of time, these beneficial differences accumulated and produced significant physical changes in the population. **With enough time, the populations branched off into new species.**

Darwin called this process *natural selection.* This evolution of the species over hundreds of millions of years was responsible for the copious, diverse life forms we see in the world today.

Darwin said that all living things were related, and this meant that *all life must be descended from a few forms, or even from a single common ancestor.* He called this process **descent with modification.**

Natural Selection

As seen in the *Paleontological Tree below*, the history of life is like a family tree. This version of the Tree was created by Ernst Haeckel, a German disciple of Darwin, in about 1879.

On the left side of the chart is the timeline showing the geologic periods, with the earliest period on the bottom. The base of the tree starts with a single life form, a primitive microorganism which Haeckel called a "moneren".

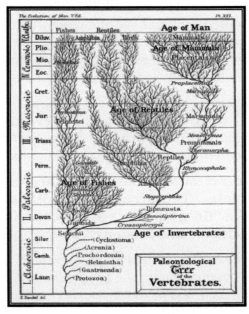

Paleontological Tree of the Vertebrates

After six branches die off, the base divides into two major limbs. This ended the Age of the Invertebrates. The first major limb on the left side leads to the Age of Fishes and finally the fish populations on the top that we see today.

The other major branch develops more slowly. But it ultimately divides into four major classes of creatures, Amphibia, Reptilia, Birds and Mammals.

The branches of the tree represent the paths taken from common ancestors that are shared amongst many different species. *The tips at the ends of the limbs represent our modern versions.* So, shown at the top of the chart is the vast diversity of life today. *On the upper right is the Age of Mammals, capped off with the Age of Man.*

To understand *natural selection,* you must first better understand *fitness.* In *evolutionary biology*[XVII], the word *"fitness" is the ability of an organism to survive and reproduce in a particular environment. An individual's fitness can be assessed by the number of offspring it has over the course of its lifetime.*

Fitness can also be measured in terms of the overall reproductive rate of a given population.

The popular phrase "survival of the fittest" was *not* coined by Darwin, but by Herbert Spencer, a British philosopher and sociologist who was intellectually active during the Victorian period, *the last half of the nineteenth century.*

The phrase is, however, **not** *adequate in evolutionary terms.* It fails to account for the critical act of reproduction. Survival alone will not lead to evolution if the creature does not contribute genes to the next generation. Perhaps a better phrase might be *"reproduction of the fittest".*

Longer survival may or may not increase fitness, because it depends on the timing and frequency of reproduction. And, most importantly, *fitness is relative to the environment.*

[XVII] **Evolutionary biology** is the subfield of biology that studies the evolutionary processes (natural selection, common descent, speciation) that produced the diversity of life on Earth.

An individual that has high fitness in one environment might have low fitness in a different environment. *Think about the fate of the dinosaurs after the environment changed so dramatically with the impact on the Earth of a major asteroid.*

In summary, *natural selection is the preservation of a functional advantage in the population that enables a species to compete better in the wild and, therefore, have a longer time to reproduce. And that reproduction does occur.*

While most scientists find the evidence for evolution to be abundant, there are some people who believe otherwise. They hold strong religious beliefs that the diversity of life that we observe was created by a higher being as described in the Bible.[79]

There are related arguments by some scientists for another concept called *"intelligent design."*[80] [81] This theory holds that certain features of the universe, and of living things, are best explained by an intelligent cause, *as opposed to an undirected process such as natural selection.* These scientists are in the minority, but they continue to seek evidence for their beliefs.[82]

The Cambrian Explosion

Looking at the fossil records, Darwin recognized that his theory had a problem. For the first four billion years of Earth's four and one half billion years existence, only the most rudimentary forms of life came into being. Most organisms were relatively simple, composed of individual cells, or small, multicellular creatures.

*Then, about 540 million years ago, over only about a 25-million-year period, **a relative flash of geologic time**, most of the modern, multicellular phyla[XVIII]* surfaced. The event was accompanied by major diversifications in other groups of organisms as well. *It was called the Cambrian Explosion.[XIX]*

In the sixth edition of **On the Origin of Species**, Darwin wrote:

[XVIII] **Phylum**- In biology, a phylum is a level of classification or taxonomic rank below kingdom and above class.

[XIX] The **Cambrian period** lasted more than 50 million years from the end of the preceding Ediacaran Period 541 million years ago to the beginning of the Ordovician Period 485 million years ago.

*"To the question why we do not find rich fossiliferous deposits belonging to these assumed earliest periods prior to the Cambrian system, **I can give no satisfactory answer.**"*

Darwin assumed, as do most evolutionary biologists today, that the fossil record is not complete, and all of the transitional forms leading from one species to another will be found. Yet, much work has been done by modern day paleontologists and there is little evidence of these transitional fossils.

One of earliest of the much more complex creatures to appear in the Cambrian period was a *trilobite*. Trilobites form one of the earliest-known groups of arthropods[XX]. They flourished and developed over 600 species at their zenith, including many with exotic exoskeletons and unique feeding strategies. Trilobites are probably most closely related to today's horseshoe crabs.

Trilobite

So, what caused the Cambrian Explosion to occur? Scientists are still mystified. There is no clear explanation for the sudden eruption of life forms. There were a number of factors that allowed the event, (necessary but not sufficient conditions), but none are considered triggering mechanisms.

First, was *the rise of oxygen in the atmosphere.* It took over two billion years for oxygen to build up to significant levels from naturally occurring phenomena. At that point, oxygen breathing animals developed.

[XX] **Arthropod** - an invertebrate animal of the large phylum *Arthropoda*, such as an insect, spider, or crustacean.

In addition, *the ozone layer surrounding the Earth appeared* around the time of the Cambrian Explosion. It shields the Earth from biologically lethal amounts of ultraviolet radiation from the sun. This protection helped the development of complex life forms on land. Previously, life was restricted to the water.

Thus, the Cambrian Explosion marks an important point in the history of life on Earth, as most of the major lineages of animals got their starts during this period and have been evolving ever since. **But the fact that the Cambrian Explosion even occurred is still a mystery, not explained by Darwin's theory, how all of these life forms developed so quickly.**

Cognitive Evolution and Animal Intelligence

The study of animal intelligence has a long history. Ever since Darwin's **On the Origin of Species** was published, scientists have attempted to understand how animals think, comparing and contrasting their styles with human thought.

The evolution of cognition is the idea that life on Earth has gone from organisms with little or no cognitive function to the wide range of cognitive abilities that we see in today's diverse creatures that populate the Earth, including humans.

Ethologists, the scientists who study animal behavior, *define animal intelligence as the combination of skills and abilities that allow animals to live in and adapt to their specific environments.* So, animal *cognition* research examines the activities that generate adaptive or flexible behavior in animal species.

Approaches to the *study* of *cognitive evolution* include focusing on *cognitive differences, and cognitive similarities between groups of the same species or two different species.*

Cognitive differences between *groups of same species* result from evolutionary **divergence**. This occurs when two groups of the same species *evolve different traits to deal with the different environments or social pressures they face.*

On the other hand, cognitive *similarities* are based on evolutionary **convergence**. [83] This occurs when two species develop *similar traits* to cope with their environments. For example, crows and octopuses have both developed a relatively high level of intelligence.

Scientists have *identified complex, cognitive capabilities that evolved in species* as they solved social challenges over a long period of time. As animals evolved, some mammals, such as baboons,[84] lions, elephants, and wolves, started living in *social groups* for various reasons. One was the division of labor to make hunting and gathering easier and more productive. Another was mutual protection from predation that was afforded by having larger size bands.

Scientists have found evidence that, while the formation of social groups was beneficial, it also presented major challenges to the individuals in the group. They had to increase their problem-solving abilities to survive in the unit. In fact, *it was the computational demands of living in large, complex societies that selected for larger brains and higher cognitive functions.*[85]

For example, the *cognitive ability to recognize individual group members* was useful to deal with many issues. One was the problem of "cheating" behavior. For instance, one member stealing another member's food supply. Individuals within the band could keep track of the rogues, then punish or exclude them from the group. This capability made the **group** more fit[XXI] than groups that did not evolve this expertise.

Another social area that required higher levels of cognition was *the formation of long lasting, interpersonal relationships.* Animals that form pair bonds and share parental responsibilities produce offspring that are more likely to survive and reproduce. This increases the fitness of these individuals and the group.

The enhanced cognitive requirements for this type of mating include the ability to differentiate individuals from their group, and to resolve social conflicts.[86] So, natural selection led to the survival and propagation of individuals with these capabilities.

Food caching is a third example of the evolution of cognition in animals based on the need to cope with their environments. This behavior, displayed by some birds and mammals, developed

---

[XXI] See Human Intelligence, Neuroscience of Intelligence, Darwin's Theory of Evolution.

because of the *requirement to find, and store food during good times to be used during times of scarcity.*[87]

A *genetic mutation* that improved fitness in some animals was *color vision.* This capability complemented the development of food caching. With the skill to identify various colors, it became easier for individuals to forage and find more fruits or other colored foods during good times.[88]

They, then, took advantage of the greater availability of a food source and used their caching strategies to store the added nutrients for later use. This whole process required the evolution of a higher-level of cognition and was a significant fitness benefit.

An example of creatures with this capability are Corvids,[XXII] including ravens, crows, rooks, jackdaws, jays, and magpies, among others. Crows, for example, have displayed incredible abilities to create and *remember the locations of hundreds of caches.*

Yet another driving force for the evolution of cognition was the *ability to use tools* and *then pass that information from one generation to the next.* Using tools provides animals with a fitness advantage, usually in the form of access to food that was previously unavailable. This provides a competitive advantage for these individuals.

Using tools was once thought to be a differentiating trait of humans. But many mammal species, as well as some birds, have demonstrated the ability to fashion and utilize tools.

For example, sea otters have been observed using a rock to break open snail shells. And some primates, as well as New Caledonian crows, have demonstrated an ability *to fashion a new tool for a specific use.*[89]

Animal Culture

It would not be efficient or effective if each generation had to re-invent the same tools. Some *animals* have shown that they are *capable of passing along their technological developments, as well as social customs, and language from one generation to the next.*[90]

[XXII] **Corvids** - There are over 120 corvid species throughout the world.

Many people believe that it is only we humans who have evolved the ability to create and pass along a culture to our young. Yet, it is surprising how many other creatures have also evolved this capability.

Apes, Cetaceans, and some birds, like corvids, have communicated information about their culture to their offspring and then later generations. For example, Orcas, or Killer Whales, in reality the largest member of the dolphin family, have been observed in the Antarctic teaching the next generation a complex hunting technique used by adults.

The Orcas also learn and teach each other local and complex *communication* techniques that are retained for many generations.

One famous example of animals passing along knowledge from one generation to the next is a group of macaque monkeys living in Japan.

A consortium of scientists went to the Japanese island of Koshima to launch a long-term study of a band of macaque monkeys that lived there.

One of the staples of the macaque diet was sweet potatoes dug from the Earth. Typically, the monkeys cleaned the potatoes of sand and dirt before they ate them by brushing them off with their hands.

One day, the scientists observed a young female clean her potato by washing it in a river. Some of her siblings saw her actions and followed suit. These macaques found a benefit in eating the cleaner food. So, the whole family was soon regularly washing their potatoes.[91]

The scientists observed carefully and found that, over the next few years, every macaque in the colony was washing their food. *So, the new behavior of using a "tool" to improve their lives was passed along to the whole macaque population on the island.*

Over the ensuing years, the scientists continued to monitor the macaques. After the generation that started the custom of washing food with water had died off, the next generation was still doing it. *The practice had become part of the macaque "culture."*

But the macaques were not alone among non-humans in appearing to have a culture. Hans Kummer was a Swiss Ethologist. He noted, independent of the macaque example, that *adaptive modifications* in animal behavior could occur. This was defined as a species using *adaptive behaviors as a result of the influence of others in the social group*. When this occurred, he agreed that *it could be considered a form of culture.*[92]

Jane Goodall, the famous researcher of chimpanzee behavior, also agrees. She has noted from numerous studies done by her and other researchers, that chimpanzee populations show widely different behavioral repertoires that can only be *attributed to cultural differences.*[93]

> *"We find that 39 different behaviour patterns, including tool usage, grooming and courtship behaviours, are customary or habitual in some communities but are absent in others where ecological explanations have been discounted.*
>
> *Among mammalian and avian species, cultural variation has previously been identified only for single behaviour patterns, such as the local dialects of songbirds. The extensive, multiple variations now documented for chimpanzees are thus without parallel. Moreover, the combined repertoire of these behaviour patterns in each chimpanzee community is itself highly distinctive, a phenomenon characteristic of human cultures but previously unrecognised in non-human species."*

Not all scientists agree that non-humans display the characteristics of what we call "culture", but Goodall and her colleagues make a good case for it.

Jane Goodall

Comparative Animal Cognition

Introduction

It has long been known that the *brain's neurons, or nerve cells, play a key role in the information processing* that goes on in the mind. These are the specialized cells that transmit and receive electrical and chemical impulses from other neurons through the dendrites and axons that interconnect the neurons (*see in this book, The Neuroscience of Intelligence, Neurons: the building blocks of the brain*).

It has been assumed that the greater the number of neurons, the greater the information processing capacity, and, therefore, the higher the cognitive abilities of the particular creature. *But, how to measure the number of neurons?*

For decades, neuroscientists have conducted studies on a wide range of animals, including humans, to try to estimate the number of neurons in their brains. Not all studies have been completely successful. Early on, it was assumed that the structure of all brains was fairly uniform, so the size of the brain of any creature was proportional to the number of neurons it had and, therefore, its cognitive ability. But this led to some problems.

Elephants and whales have exceptionally large brains compared to humans. A sperm whale's brain, for example, weighs about 17 pounds, while a human brain is about three pounds. While whales are relatively intelligent creatures, their cognitive abilities are not thought to be even close to humans. So, pure brain size is not the answer.

The next step was to correct for the size of the brain compared to the mass of the body it controlled. The idea was that the higher the brain-to-body mass ratio of an individual, the more brain capacity left over for cognitive functions.

So, to test the theory, neuroscientists compared brain mass to body mass of a large number of animals, as shown in the Figure on the next page. The results are plotted on a log-log scale[XXIII] in

---

[XXIII] **A logarithmic scale** (or log scale) is a way of displaying numerical data over a very wide range of values in a compact way. The scale is non-order to fit small creatures with small brains and large ones with large brains on to the same chart. The line that runs from the bottom-left to the top-right represents the "average"[XXIV] of all the data points. The further Up and to the Left of the line, the larger the brain mass to body mass ratio.

The data points are generally in the direction that is intuitively correct. Whales, for example, have relatively high ratios, but lower than humans. Primates, such as chimpanzees, gorillas, and orangutans, are closer to the human ratio, but as expected, still lower. *One problem, however, was that small rodents, like mice, had brain/body ratios like those of humans (both are 1/40). If their brains had a similar structure to humans, they should be much more intelligent than they appear.*

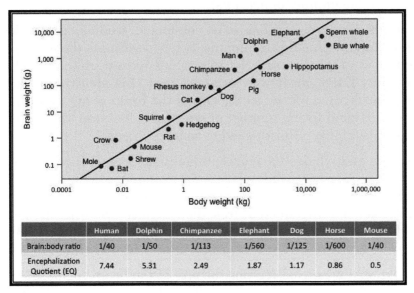

	Human	Dolphin	Chimpanzee	Elephant	Dog	Horse	Mouse
Brain:body ratio	1/40	1/50	1/113	1/560	1/125	1/600	1/40
Encephalization Quotient (EQ)	7.44	5.31	2.49	1.87	1.17	0.86	0.5

Brain Mass to Body Mass

linear. Typically, moving one unit of distance along the scale means the value has been multiplied by 10. So the first division is 10, the second is 100 and the third 1000. On such a chart, an exponential function displays as a straight line.

[XXIV] **A line of best fit**, determined by linear regression, is also called a trend line. It is a straight line drawn on a graph that best represents the data on the plot.

The conclusion was that brain-to-body mass ratio, and a later refinement of that measure, called the Encephalization Quotient[XXV], was a better representation of relative cognitive ability than just brain size, *but still imprecise for many of the creatures studied.*

It became clear that, *what was needed was a better methodology to directly measure the number of neurons in a brain rather than considering the weight of the brain as a proxy for the number of neurons.* Stereology[XXVI] is a standard tool in the biosciences and has been used for many years to count the neurons in the brain, one small section at a time.

One such device is the *Optical Fractionator.* Because the brain's neurons are *not evenly distributed throughout its structure*, this method turned out to be terribly slow and subject to significant errors. As a result, *no reliably accurate counts of neurons in animal brains emerged as a result of using this process.*

In about 2005, a young Brazilian neuroscientist, made a *breakthrough* in the ability to count neurons quickly and accurately. *Dr. Suzana Herculano-Houzel* developed a method to strip out the nuclei of neurons from a brain sample using detergent. Then she homogenized the solution and used standard, electro-optical instruments to count the neurons quickly and easily.

Officially, her method is called an Isotropic Fractionator. She prefers to call her process by her informal, but accurate, name, "brain soup". Since then, she and her colleagues from Vanderbilt University, where she now works, have conducted a wide range of

---

[XXV] **The encephalization quotient (EQ)**, is a relative brain size measure that is defined as the ratio between observed brain mass compared to the *predicted* brain mass for an animal of a given size and structure.

[XXVI] Stereology is the three-dimensional interpretation of two-dimensional cross sections of materials or tissues. It provides practical techniques for extracting quantitative information about a three-dimensional material from measurements made on two-dimensional planar sections of the material.

studies on comparative brain anatomies across many species, and reached some remarkably interesting conclusions.[94]

Dr. Suzana Herculano-Houzel

One of her major findings is that our human brains contain an average of 86 billion total neurons, many more than any other species. Previously, neuroscientists estimated the number to be about 100 billion, but no one had ever actually measured the actual number because of the limitations of the technology.

She also concluded that the *human brain is not special.* It is *just a scaled-up primate brain.* So as brain size increases so does the neuronal count of that primate brain.

Then, the question arises, *"why did we evolve with this larger brain while other primates did not?"* According to Herculano-Houzel, it is *because humans started cooking their food.*

Primate brains require a significant fraction of the body's energy to operate properly. A human brain, for example, only accounts for about 2% of the body's weight. Yet, it requires a substantial fraction, 25%, of the calories consumed by an individual.

Gorillas and orangutans have much larger bodies than humans. And they consume a low calorie, vegetarian diet. During their evolution, their diet did not provide enough calories to support both a large body *and* a large brain. Therefore, they evolved with a *smaller brain* for their body size, with proportionally fewer neurons.

Humans, on the other hand, developed cooked food. Our "cave man" ancestors, the Homo erectus people, started cooking their food about 500 million years ago. This provided them with much more densely packed calories in the food they consumed, and they could support a much larger brain for their smaller sized body. Evolution did the rest.

Dr. Herculano-Houzel's work also uncovered the reason why the brain-to-body-mass ratio of mice, and for most rodents, was like humans, yet the creatures were without our level of intelligence. *The answer turned out to be the structure of the neurons in the rodent brain and scaling factors as brain size increased.*

While studying rodent brains, Dr. Herculano-Houzel found that the *average size of an individual neuron in a rodent brain* **increased with body size**. So, *the brain size for larger rodents was bigger than expected (by scaling) for the number of neurons it contained.*

In primates, like us, on the other hand, the neuron size remained the same as the body and brain size increased. The result was that primate brains gained many more neurons than rodents with an increase in brain size.

Looking at primate brain structure in more detail, the **cerebral cortex is the outer layer**. It is grey and has many folds called gyri. This image is what you usually think of when you picture a brain[XXVII].

The cortex is the part of our brain that is responsible for our awareness, thought, memory and language. In other species, it is also the seat of higher thought processes.

To compare intelligence between species, *scientists agree we should focus on the number of neurons contained in* **the cerebral cortex,** *not the whole brain.* As shown in the Figure on the next page, based on data using the "brain soup" methodology (isotopic fractionator), *humans have an average of 16 billion neurons in their cerebral cortex. Gorillas and Orangutans,* among our closest relatives in the animal world, *have about 9 billion each, or 44% fewer than humans.* While *African elephants,* with much larger

---

[XXVII] In this book, see *Human Intelligence, II. The Neuroscience of Intelligence, A. Neuroanatomy, The Structure of the Brain.*

total brain sizes, are measured at close to *6 billion, or 63% less cortical neurons than a human.*

Cetaceans, such as whales and dolphins, are also among the creatures with the highest intelligence potential with *cerebral cortex neurons in the 3 to 6 billion range.* The number is about the same for chimpanzees, but on the higher end of the spectrum.

Common name	Average number Neurons in Cerebral Cortex (millions)
Golden Hamster	17
Eastern Mole	24
Brown Rat	31
Guinea Pig	44
Eastern Grey Squirrel	77
Common Blackbird	136
Common Starling	226
House Cat	250
Brown Bear	251
Capybara	306
Red Fox	355
Racoon	453
Lion	545
Golden Retriever	627
Grey Parrot	850
German Shepherd	885
Raven	1,204
Rhesus Macaque	1,710
Pilot Whale	3,000 (est.)
African Elephant	5,600
Gorilla/Orangutan	9,000
Human	16,340

Note: All measurements made with an isotropic fractionator, or the "brain soup" method

So, which creature is more intelligent, your dog or your cat? Dogs were found to have on average *for all sizes,* about 525 million cortical neurons, while cats have about 250 million. That is, cats have less than half of the neurons in the average dog's cortex. However, as an overall comparative, each pet has less than 6% of a primate's neurons, so they are both low on the mammalian intelligence scale. *But the dog wins handily in the comparison with cats.*

The g-Factor in Animals

Psychologist Charles Spearman found in 1927 that people who do well on one of the cognitive exams, e.g. vocabulary, which make up the battery of an IQ test, also tend to do well on all the other components. He postulated, based on this observation, that *there is an underlying, **general intelligence factor** in humans that he termed, "g".*[95] XXVIII

Technically, "g" is a variable that summarizes the correlations of performance between the different, specific cognitive tasks that, together, make up an IQ exam.

The standard approach to calculating the "g" factor is to use a statistical technique called *exploratory factor analysis.*[96] But there is no need to calculate an individual's "g". It turns out that the full-scale IQ score from the test battery has been *found to be highly correlated with "g" factor scores, **so the IQ score can be regarded as an estimate of "g" for that individual.***

The g-factor in humans is widely accepted as being accurate and useful. As a result, many scientists have explored the idea that a similar, g-factor of intelligence exists for animals. To study this phenomenon, they observed animals in natural settings, or on behavioral tasks in experimental surroundings, to develop data on their level of intelligence.

The observed activities included such things as innovation and problem solving. This type of behavior involves watching the animal develop novel solutions to everyday problems. Or it may be the observation of the animal's use of tools to achieve a goal.

XXVIII See in this book, *Human Intelligence, How intelligence is measured.*

One example is chimpanzees using a stick to fish for termites. Such observations provide valid measures without the need for experimentation.

Like an IQ examination, a comprehensive assessment of an animal's intelligence often includes a battery of tests involving several sorts of behaviors.[97] An example was a study done on Border Collies. They were the subjects of a study to determine if dogs had a g-factor.[98] (See the Section *"How intelligence is measured"*).

Their most important finding dealt with how well the dog's score on one of the problems predicted its scores on the other tests. It is important to note that the other two tests involved quite different mental abilities.

When the researchers did the statistical analysis, they found that dogs which were good on the first problem were also good on the other two classes of problems. While dogs who were not so good on the first problem did considerably less well on the other tests. ***This is a finding that is similar to Spearman's results with humans.***

Arden and Adams concluded that the result provides strong evidence for the idea of a g-factor in intelligence with smart dogs being generally proficient at everything and not so smart dogs doing generally more poorly on most other measures. *Therefore, they concluded, that dogs do indeed have a "g" factor.*

In fact, thousands of such studies have been done over the years and published in peer-reviewed journals. A methodology was needed to bring all of the findings together. *Meta-analysis is such a tool.* It is a statistical examination of the data from a number of independent studies of the same subject, in order to determine overall trends.

This type of analysis was done for *four thousand academic papers on primate behavior*.[99] The researchers were looking for instances of innovation, social learning, tool use, and extractive foraging. The idea was to investigate the components of these behaviors in sixty-two different species of primates.

*The analysis of those studies revealed **a single factor** explaining 47% of the variance onto which the cognitive measures loaded. This* *suggests that non-human primates, as a whole, also have a g- factor similar to that observed in humans.*

There is *not yet, however, a universal acceptance by scientists* of a g-factor in animals.[100]   An article published in the **Proceedings of the Royal Society** in 2020 by Marc-Antoine Poirier of the University of Ottawa, contained a review and meta-analysis of *"g"* in non-humans.   He based his analysis on 49 published articles.   His work found that the *average correlation coefficient between cognitive tasks was only 0.18*.[XXIX]   This result, Poirier concluded, provides very weak support for the idea of a g-factor in animals.

The Great Apes

It is general knowledge that Great Apes, including gorillas, chimpanzees, bonobos, and orangutans, are among the most intelligent of all the animals below man.   And that fact is confirmed by the many scientific studies of these remarkable creatures.

It was once thought that only humans could make logical inferences. Animal cognition researcher, *Josep Call, however, has shown that apes are also able to draw such conclusions.*[101]

Dr. Josep Call

[XXIX] **The correlation coefficient**, *r*, tells us about the strength and direction of the linear relationship between *x* and *y*.   An r of, say, +0.7 would be considered to be significant, whereas +0.18 would be considered to be weak.

In experiments done by Call at the *Max Planck Institute for Evolutionary Biology*, a group of bonobos, gorillas, chimpanzees, and orangutans were presented with two cups. The animals could see that one cup was empty and one contained a favorite food.

Then the researchers shook the two cups in front of the apes. Only the baited cup made a rattling sound. Upon repeated demonstrations the apes made that appropriate association.

Later, the subjects were shown the two cups without seeing which one had food. They were allowed, however, to see the cups being shaken. First, the one with food inside rattled and they could hear it. Then the other, empty cup was also shaken in front of them, but made no sound.

The apes were then allowed to pick one of the cups. The majority of the time, they chose correctly and picked the baited cup that had made the noise.

In the next step, the two cups were presented, but *only the empty cup was shaken* in front of the subject, and, of course, it made no noise. The subject was then allowed to pick one of the cups, and they still correctly chose the one with the food.

This suggests that the subjects made an inference by exclusion since they chose correctly without having heard the food rattle in the other cup. They knew there was no food in the cup without the rattle, so they selected the other one and were rewarded.

Control tests were conducted and showed that the subjects were not more attracted to noisy cups, or avoided cups that were shaken and noiseless, or learned to use auditory information as a cue during the study.

A *major conclusion by the researchers was that the apes understood that the food caused the rattling noise, not simply that the noise was associated with the food.*

On a personal note, whenever I visit a city for the first time, I like to check-out their zoo. I was between appointments in Basel, Switzerland and decided to visit their wildlife park. There were hardly any other visitors on that day, so I had the place virtually to myself. I vividly remember my encounter with one of that zoo's orangutans.

He looked relatively young, a teenager perhaps. He was sitting near the bars of his cage when I approached. We were just a few feet apart. What impressed me was the way he looked back at me. He had large, soulful, dark eyes that stared into mine. We just remained there, watching each other for several minutes. Neither of us made a sound, but I could feel the deep intelligence behind his stare. I was extremely impressed by my fellow primate's "human like" nature. He seemed sad, as I imagine I would be if I were locked into his cage.

Orangutan

I hope he is much happier now. In 2012, the Basel Zoo opened its new, Geigy Enclosure for apes. It is a spacious outdoor facility with many features to keep apes occupied.

Orangutans are among the most intelligent of the Great Apes. They are only found across the world in the rainforests of Borneo and Sumatra. Borneo covers an area slightly larger than Texas, while Sumatra is a little smaller.

Both are Southeast Asia islands that are situated on the Equator. Because of their hot, wet climates, they have the world's most diverse, dense rain forests, perfect habitat for large, tree dwelling, apes.

Many scientists are interested in studying orangutan cognition. One such study is underway in the zoo of Atlanta, Georgia. The scientists have built a computer game operated by a touchscreen and placed it in the orangutan habitat. In one game, the orangutans match identical photographs, or orangutan sounds with photos of the animals. Correct answers are rewarded with food pellets.

The object of the experiment is to test the animals' memory, reasoning, and learning. The program provides a significant amount of data for researchers at the zoo and Atlanta's Center for Behavioral Neuroscience, which is a partner in the project.

"The data will help researchers learn about the socializing patterns of these creatures, such as whether they mimic others or learn behavior from scratch through trial and error", said Elliott Albers who works with the Center for Behavioral Neuroscience.[102]

In other studies, researchers have found a deep-rooted intellect in orangutans, capable of serious problem solving and innovation. This includes toolmaking, and recognition of self in a mirror.

On the social side, Orangutans are capable of deception, forming alliances and coalitions, having empathy for another orangutan, and consoling that individual when it is feeling badly. They have also been observed teaching one another and passing along "culture", such as the ability to make and use tools.

In a 1997 study, two captive, adult orangutans were tested with *the cooperative pulling model.* This test required the two apes to coordinate their actions in order to obtain the food reward. Without any training, the orangutans succeeded in working together to pull down an object to get at the food underneath.

They accomplished the feat in the first session. Over the course of thirty sessions, *the apes succeeded more quickly with time, having learned to better coordinate their actions.*[103]

In another study, conducted in 2008, involving two orangutans at the Leipzig Zoo, it was demonstrated that orangutans can use "calculated reciprocity" in a transaction. This involves the primates weighing the costs and benefits of gift exchanges between each other and keeping track of the exchanges over time. *Orangutans are the first, documented, nonhuman species to demonstrate that they can accomplish such a feat.*[104]

Chimpanzees and Bonobos are closely related members of the chimpanzee family. Both are human's closest relatives in the animal kingdom, having diverged from a common ancestor about four and a half million years ago. Bonobos and Chimpanzees share close to 98% of their genome in common with humans. This implies that their genomes are more similar to that of humans than they are to even those of gorillas.

Like orangutans, they are remarkably intelligent. One indication is that they make extensive use of tools. For example, they use grass stalks, vines, or branches to "fish" for termites in otherwise unreachable places in trees or mounds.

Chimpanzees have been observed using stones or hard wood as hammers or anvils to crack open tough nuts. These apes also use different tools in succession as a "tool set." For example, in the West Africa country of Guinea, chimps have developed tools to get at hard-to-reach water. They push leafy sponges into hollows of trees containing the liquid and then withdraw the wet sponges by using sticks.

In the wild, they communicate extensively with each other. These communications take the form of facial expressions, gestures, and a large array of vocalizations that include screams, hoots, grunts, and roars.

But, in captivity they have been taught to communicate with humans through sign language, or language based on the display of tokens or pictorial symbols.

Washoe was a female chimpanzee who was the first non-human to learn, and then communicate using American Sign Language (ASL). The project started in 1967 at the University of Nevada, Reno. The effort was headed by Drs. Allen and Beatrix Gardner who *"adopted" Washoe, then almost two years old, and raised her as a human child.* The goal with raising her as a human was to attempt to satisfy her psychological need for companionship.

Washoe with the Gardners

Previous studies with chimpanzees attempted to teach them to literally speak a human language, but these efforts failed. Chimpanzees do not have the necessary vocal ability to imitate human speech. So, the Gardners wanted to try teaching Washoe sign language.

Eventually, they found that Washoe could pick up ASL gestures without direct instruction. Instead, she learned by observing humans signing amongst themselves.

For example, the scientists signed "toothbrush" to each other while they brushed their teeth in front of Washoe. On a later occasion Washoe reacted to the sight of a toothbrush by spontaneously producing the correct sign, *thereby showing that she had in fact previously learned the ASL sign.*

When she was five years old, Washoe moved to the University of Oklahoma's Institute of Primate Studies in Norman, Oklahoma, under the care of Drs. Roger and Deborah Fouts.

Over the course of her life, Washoe learned approximately 350 words of sign language.[105] For researchers to consider that Washoe had learned a sign, she had to use it spontaneously and appropriately for 14 consecutive days.[106]

When other researchers tried to replicate the language acquisition results of Washoe, they were not nearly as successful. One well-known such project was run by Drs. Herbert Terrace and Thomas Bever of Columbia University. Their chimpanzee subject was *Nim*

Chimpsky, named after the famous, MIT linguist, Noam Chomsky.[107]

Nim Chimpsky

Nim was trained successfully to use 125 signs. Terrace and his colleagues, however, concluded that the chimpanzee did not show any human-like use of language. Nim's use of language was instead strictly pragmatic. It was a means of obtaining an outcome. *They concluded that there was nothing Nim was taught that could not equally well be taught to a pigeon using the principles of operant conditioning.*

The researchers therefore questioned claims made on behalf of Washoe and argued that the apparently impressive results may be nothing more than the consequence of a "Clever Hans" effect.[xxx]

Other great apes have also learned sign language including orangutans and gorillas. One well known example is Koko the female, western, lowland gorilla born in the San Francisco zoo. She lived most of her life in Woodside, California, at The Gorilla Foundation 's preserve in the Santa Cruz Mountains.

[xxx] **The Clever Hans effect** - It is a term, named after a trick horse, used in psychology to describe the situation where an animal, or a person, senses what the "trainer" wants them to do, *even though they are not deliberately being given signals*. It is important to take this effect into account when testing animals' intelligence.

Her instructor and caregiver was Dr. Francine Patterson, an Animal Psychologist. Patterson reported that Koko had an active vocabulary of more than 1,000 signs. This puts Koko's vocabulary at the same level as a three-year-old human.[108]

Koko the gorilla with Dr. Patterson

Other projects that attempted to teach sign language to non-human primates, like Washoe, purposely did not expose them to spoken language. On the other hand, *Patterson simultaneously exposed Koko to spoken English from an early age, along with the signs.* It was reported that Koko understood approximately 2,000 words of spoken English, in addition to the signs.

Koko's life and learning process has been described by Patterson and various collaborators in books, peer-reviewed scientific articles, and on a website.[109]

As with the other great-ape language experiments, *it was greatly debated whether Koko mastered and demonstrated language through the use of signs.* However, she did manage to score between 70 and 90 on various IQ scales.

Dr. Mary Lee Jensvold of Central Washington University is Director of the Chimpanzee and Human Communication Institute. She has also worked extensively with Washoe the chimpanzee, so Dr. Jensvold is an expert in great apes communicating with sign language. *She claimed that Koko did indeed use language the same way people do.*[110]

Cetaceans

I am fortunate to live along the coast of the Atlantic Ocean in Gloucester, Massachusetts. I have sailed the waters around Gloucester for many years, so I have had close encounters with Cetaceans on numerous occasions, never dangerous, always pleasant. I am constantly impressed with their behaviors.

"Whale watch" boats are a summertime staple of the Gloucester fleet. I have been onboard as they head out of the harbor and sail Southeast for about seventeen miles to an area called Stellwagen Bank. It is a relatively shallow (one hundred feet), underwater plateau. This happens to be an area where krill are plentiful. They are the small, shrimplike, planktonic crustaceans that are the major food source for baleen whales, like humpbacks.

It is not a common event on such a whale watch, but once in a while it is exciting to see a huge humpback whale leave the water in a flying breach and make an enormous splash. The resulting sound travels long distances underwater. Some scientists say it is one of the whale's modes of communication.

Breaching!
Oil painting by Anthony J. Marolda

Cetaceans consist of *whales, dolphins, and porpoises.* They divide into toothed whales, like belugas and sperm whales, and baleen whales[XXXI], like right whales and humpbacks. All whales are renowned for being highly intelligent and have been the subjects of numerous studies.

Dolphins and porpoises both have teeth and look similar, but with some clear differences. Dolphins are sleek and have a narrow snout, while porpoises are portly and have a flat snout.

Dolphin

Porpoise

Porpoises are found primarily in the Pacific Ocean while dolphins are found all over the world. More importantly, dolphins are much more social creatures than porpoises. As a result, they make a wide variety of sounds useful for communication. Porpoises tend to be quieter reflecting their more solitary natures.

The evolution of Cetaceans is believed to have begun about fifty-five million years ago on the subcontinent of India. They began as Indohyus, a relatively small, pig-like, hoofed, land mammal with a small brain. But, over that time, they developed into the highly adapted mammals we see today.

[XXXI] Baleen is a filter-feeding system inside the mouths of baleen whales. Baleen acts as a sieve, catching tiny organisms in sea water, which the whale consumes as food

Indohyus

These adaptations to the ocean environment include some fascinating characteristics such as echolocation. This is an ability found in toothed whales as an aid in hunting their prey. It is the location of fish by reflected sound similar to the ability of bats in locating flying insects.

Cetaceans also evolved intricate auditory and communicative capabilities. This expertise helps with maintaining their complex social organizations. They live, travel, and hunt in groups called pods. This allows them to better survive by hunting cooperatively as a band.

Scientists who study whale behaviors are careful not to anthropomorphize them. Yet, the more we learn about whales, the more we see that they are similar to us in many ways.[111]

Living in pods allows them to pass along their cultures to the younger members of the group. The cetacean cultures include communication protocols and hunting techniques. Some of this knowledge has been observed to stay within the pod, so that each group may have their own unique culture. There can, therefore, be large variations between pods in behaviors like communication, how they care for their young, and their hunting techniques.

All of these developmental factors contributed to the evolution in Cetaceans of large, complex brains with a high number of cortical neurons. And, therefore, a high degree of intelligence.

Although Cetaceans have not shared a common ancestor with primates for over ninety million years, *they possess a set of cognitive attributes that are remarkably convergent with those of many primates, including great apes and humans.*[112] Cetacean brains and primate brains are considered to be alternative, evolutionary routes to the same, significant level of neurobiological and cognitive complexity.

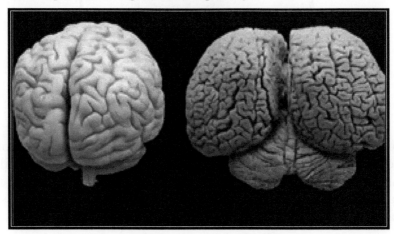

Human brain vs Dolphin Brain

As can be seen above, the modern Cetacean brain, *on the right*, is among the largest in terms of the number of cortical neurons, absolute size and in relation to body size, as measured by the Encephalization Quotient (EQ).[XXXII]

More specifically, the evolution of the dolphin brain was the result of the different requirements for adaptation to the aquatic environment. *This included their highly efficient and effective echolocation system.* Having this capability facilitated a successful feeding strategy for the dolphin to support its increased metabolic requirements for its larger brain.

Included with the total brain size increase of the Cetaceans was a significant increase of the cerebral cortex and, therefore, the number of neurons and synapses. This, in turn, enabled their high level of cognitive capabilities, as well as their precise and rapid sound processing skills.

[XXXII] See the Introduction to this section, "Comparative Animal Cognition".

As shown in the Figure above, the high degree of a cetacean's *cortical gyrification*, the characteristic folds of the cerebral cortex, and the resulting expansive surface area, is unsurpassed among mammals, including humans.[113].

On the other hand, *primate* brains *feature large frontal lobes. No similar region is found in the cetacean brain.* Instead, *cerebral enlargement in Cetaceans occurs in the parietal and temporal regions. These are the areas* associated with the animals' auditory capabilities. This gives them the brain power required for their unique echolocation and communication abilities.

The total number of neurons in the dolphin's cortex, the area of the brain involved in higher cognitive processes, is *comparable to that of the chimpanzee, or about 6 billion.* This compares to sixteen billion neurons found in a human's cortex, which amounts to about 166% larger than the dolphin.

Interestingly, recent studies of the baleen whale brain, such as the humpback, demonstrate that the cortex of these large mammals, although different in cytoarchitecture[XXXIII] than a dolphin, are comparably large in absolute terms. It is estimated that whales have approximately the same number of cortical neurons as dolphins.[114]

Significant research has gone into assessing the problem-solving skills of Cetaceans. These studies demonstrate that, in spite of the significant differences in cytoarchitecture, humans and Cetaceans share important cognitive abilities. These include the ability to use a symbolic language, coordinate social behaviors and demonstrate an awareness of one's own actions, but at obviously different levels.[115]

Some of the evidence for the high level of dolphin cognition is anecdotal but is still interesting to note. For example, dolphin researcher Pieter Arend Folkens tells this story:[116]

[XXXIII] Cytoarchitecture: the arrangement of cells in specific areas of the cerebral cortex associated with particular functions, such as hearing, vision, cognitive functions, etc.

"Since trash can be dangerous to dolphins if ingested, some of the animals at Marine World Africa USA[XXXIV] were trained to retrieve the trash and return it to the trainer for a reinforcement reward.

"A trainer would come out onto the floating stage and a dolphin would perform a tail stand with a piece of trash in its mouth. The trainer would then reward the dolphin with a bit of fish.

"One day the lead trainer went through the routine only to notice that the dolphin kept coming back with a piece of trash even though the tank appeared clean. The trainer asked a colleague to go below to the engineer's port to observe what the dolphin was doing when a trainer came out on the float.

"The scam was revealed! This dolphin had established a savings account of sorts. He collected all the trash and stuffed it in a bag wedged in a corner of the tank near the intake of the filtering system. In there was paper, rope, and all sorts of trash. The amazing thing is that when he went to the bank, he did not simply take a piece, rather he would tear a bit off to maximize the return.

"This behavior is particularly interesting because it shows that the dolphin had a sense of the future and delayed gratification. He had enough presence to realize that a big piece of trash got the same reward as a small piece, so why not deliver only small pieces to keep the extra food coming? **He in effect had trained the humans."**

Orcas, or "Killer Whales", are top predators, *but they are not whales.* They are actually the largest member of the dolphin family. *They are recognized as probably the most intelligent of all Cetaceans.* They have the second-biggest brains among all ocean mammals, weighing as much as 15 pounds. The largest brain, at 17 pounds, belongs to the Sperm whale. But that animal is sixty

---

[XXXIV] Marine World and Africa USA were animal parks located in Redwood Shores, California. They merged in 1972. They moved and are today Six Flags Discovery Kingdom.

feet long, twice the length of an Orca. *So, an Orca's Encephalization Quotient is significantly greater than the Sperm whale and is similar to that of a chimpanzee.*

Orca

As can be seen in the above Figure, Orcas have a distinctive marking pattern, their back is black, they have a white chest and sides, and a white patch above and behind their eye.

Another strong characteristic of Orcas is that they have, and maintain, robust cultures. Different sub-groups may have very specific skills, such as hunting techniques or dialects, which they teach and pass along to their young. These skills will stay within specific groups for generations. One researcher even said that the only species more "cultural" than Orcas are humans.[117]

One particular use of Orca brain power is to have a sophisticated means of communicating with each other. Basically, they use three types of sounds, clicks, whistles and pulsed calls, to make up their particular dialect. Another group of Orcas use the same sounds but may have a completely different dialect, as different as English is from Japanese.[118]

Orcas are particularly good at listening and imitating others. This is a capability useful in passing along a culture. A study found that Orcas can even learn to speak "dolphin".[119] An Orca which spent considerable time with bottlenose dolphins started to make more clicking sounds, replicating those used by the dolphins in their communications within their pod.

The mirror recognition test is a well-used, animal behavior exam. Its purpose is to learn if animals are able to recognize themselves, and, therefore, if they are self-aware. Orcas are one of a very few animals that have passed this test. Other animals that have also passed the test are the Great Apes, Asian Elephants and Magpies, a Corvid.[120]

Elephants

Elephants are the largest, existing land animal. There are three species, the African bush elephant, the African forest elephant, and the Asian elephant. The Africans and the Asians look similar but have a few different features. They are both large, with the African being somewhat bigger than the Asian, greyish brown in color, mostly hairless with tusks and a long proboscis or trunk. The prehensile trunk is used for breathing, bringing food and water to the mouth, and grasping as well as manipulating objects.

But the African elephants have larger ears and concave backs, whereas Asian elephants have smaller ears, and convex or level backs.

Asian elephants have a larger, higher forehead. They are the species used by circuses in the United States, so they are what most people think of when considering an elephant.

African elephant (left) and Asian elephant

Asian elephants have been used for work by humans for centuries and, in most cases, have been well cared for. They are more docile than African elephants who have been heavily hunted by humans over the same period of time.

African elephants possess long memories and are more wary of humans, which may account for their more aggressive behavior.

They are, therefore, not used by humans for work in the same way that the Asian elephants are employed.

Both species, however, are known for their high intelligence. The first indication of a similar level of intelligence is the number of neurons in the cerebral cortex as determined by the isotopic fractionation methodology (Herculano-Houzel's "brain soup" technique). As seen in the table found in *Comparative animal cognition, Introduction,* Elephants are near the top of the animal kingdom in this regard. They have close to six billion neurons in their cerebral cortex which is comparable to Orcas. They are only exceeded by the Great Apes at nine billion cortical neurons and humans at sixteen billion.

Elephants are also among the few mammals who have a significant number of von Economo, or spindle neurons, in their brains.[XXXV] The other mammals include humans, cetaceans, and great apes, each of which has a relatively large brain.

Although they are rare in comparison to other neurons, von Economo neurons are abundant, and comparatively large in these creatures. *They are believed to allow rapid communication of information across areas of their relatively large brains.* A necessary condition for high intelligence.

The fact that von Economo neurons appear in *only distantly related* clades[XXXVI] (e.g. humans vs. cetaceans) suggests that they *represent convergent evolution* – specifically, as an adaptation to accommodate the increasing size of these distantly-related, but large brains.

There is considerable evidence from observation and studies that elephants are indeed among the brightest of creatures. For example, elephants are able to learn new facts and behaviors,

---

[XXXV] Von Economo neurons were first identified approximately one hundred and forty years ago by the Russian scientist Vladimir Betz. Later, they were named after the anatomist, Constantin von Economo. Von Economo, or spindle, neurons are found in the brains of humans, great apes, whales, dolphins , and elephants.

[XXXVI] A *clade*, also known as a natural group, is a group of animals composed of those with a common ancestor and all its lineal descendants - on a phylogenetic tree.

mimic sounds that they hear, self-medicate, play with a sense of humor, perform artistic activities, use tools, and display compassion and self-awareness. It has also been shown that elephants are capable of a range of emotions, including joy, playfulness, grief, and mourning.

Perhaps one of the clearest indications of intelligence in elephants is their ability to use tools to solve problems or accomplish necessary tasks. An important prerequisite for an elephant's use of tools is their prehensile trunk, with its finger-like tip. The trunk is capable of manipulative movements similar to those performed by primates with their fingers and thumb.

For example, Asian elephants in India occasionally encounter an electrified fence designed to keep them away from certain land areas. If they are determined to get through, they find a way. They have been observed using their trunks to pick up large rocks or logs and then dropping the object onto the fence to break it. Then they use their tusks to clear the wires and create a safe passageway for the herd to pass through.

Elephants have also been observed using their trunks to pick up leafy branches to be used as fly switches, and short, sturdy sticks to de-tick themselves. It has been noted by some researchers that Asian elephants actually modify long branches to make them the ideal length for fly switching or de-ticking.[121]

Elephants have been found to have an amazing ability to identify humans who may pose a threat to them merely by hearing their voices. Professor Karen McComb, and her team, are part of the *Mammal Vocal Communication and Cognition Research* department of the School of Psychology at the University of Sussex in the U.K. She works with African Elephants at the Amboseli National Park in Nairobi, Kenya. The researchers presented their findings on elephants' ability to identify human voices in a peer reviewed journal.

Professor Karen McComb

Their hypothesis is that *animals can accrue direct fitness benefits by accurately classifying a predatory threat using auditory information.* Human predators, however, present a particular cognitive challenge for elephants since different human subgroups pose radically different levels of danger.

For example, McComb and her team found in their experiments that the elephants could distinguish between the recorded voices of men of different ethnic tribes, and who pose different levels of threats.

The Maasai tribe in Kenya have come into conflict with, and hunted, elephants for centuries, whereas the men of the Kamba tribe have not. The elephants apparently learned to distinguish the differences in sound between members of the two tribes. They could distinguish the Maasai men's' voices, who posed a danger, from the voices of the Kamba men who did not.

They could also distinguish the voices of *Maasai women and children* who also *did not* pose a threat. [122]

> *"Our results demonstrate that elephants can reliably discriminate between two different ethnic groups that differ in the level of threat they represent, significantly increasing their probability of defensive bunching and investigative smelling following playbacks of Maasai voices. Moreover, these responses were specific to the sex and age of Maasai presented, with the voices of Maasai*

women and boys, subcategories that would generally pose little threat, significantly less likely to produce these behavioral responses.

Considering the long history and often pervasive predatory threat associated with humans across the globe, it is likely that abilities to precisely identify dangerous subcategories of humans on the basis of subtle voice characteristics could have been selected for in other cognitively advanced animal species."

Another clear indication of Elephants high level of intelligence is the complexity of their social interactions. They have one of the most closely knit societies of any living species. Elephant families can only be separated by death or capture.

There are many observations of elephants taking care of other individuals in some type of difficulty, or even helping other species including humans.[123] It is *not common* in the animal kingdom for uninvolved bystanders to provide consolation to others in distress. *Great Apes and corvids are among the few species to display such behavior.* What is common is that *these species are among the most intelligent.*

In a study carried out at an elephant camp in Thailand, Plotkin, et al, found that uninvolved elephants connected empathetically with other individuals following an event that caused stress for that creature. The elephants showed their concern for the distraught individual through directed, physical contact and vocal communication with that animal.[124]

The level of contact was much higher during the stress event than in control periods where no traumatic situation had occurred. The elephants' consolation behavior was similar to the responses by apes in similar situations. Plotkin theorizes that this response is based on *convergent evolution of empathic capacities in the two clades.*

Avians

Dr. Suzana Herculano-Houzel and her team at Vanderbilt University have also done considerable work on avian brains.[125]

XXXVII *She found that birds are remarkably intelligent.* They have relatively small brains, but their brain structure is different from mammals. The neurons are smaller in scale and more densely packed. As a result, Corvids (ravens and crows) and some parrots are capable of cognitive feats comparable to those of great apes. Herculano-Houzel said:

> *"How do birds achieve impressive cognitive prowess with walnut-sized brains? We investigated the cellular composition of the brains of 28 avian species, uncovering a straightforward solution to the puzzle: brains of songbirds and parrots contain very large numbers of neurons, at neuronal densities considerably exceeding those found in mammals. Because these "extra" neurons are predominantly located in the forebrain, large parrots and corvids have the same or greater forebrain neuron counts as monkeys with much larger brains. Avian brains thus have the potential to provide much higher "cognitive power" per unit mass than do mammalian brains."*

Corvids

Corvids are any of the family (Corvidae) of stout-billed, perching birds including the crows, jays, magpies, and the raven. As a group, they are known to be especially intelligent.

People have been watching crows for a long time and reporting on their remarkable behaviors. For example, crows in Japan, as reported by naturalist Sir David Attenborough on the BBC "Wildlife" show[126], have learned that they can drop hard walnuts on a roadway and wait for a car to run it over, so they can get the nut inside. But there was a problem.

XXXVII See Comparative animal cognition, Introduction, Table showing number of neurons in cerebral cortex by species.

Crow eating a nut on road

The crows realized that they could be hit by an oncoming car as they tried to retrieve the food. So, they figured a way around that danger. They dropped the nuts onto a crosswalk associated with a traffic light and waited for the nuts to be run over there. Later, when cars stopped for the traffic light to let pedestrians pass on the crosswalk, the black birds were able to swoop in to get their reward without any danger of being hit.

Beyond anecdotal evidence, scientists have done many scientific studies on the depth of crow intelligence. One measure of intelligence in any creature is the ability to use tools to solve problems. And the more complex and multi-stepped the task is, the greater the intelligence required.

A research team, headed by Dr. Alex Taylor, at the University of Auckland in New Zealand tested a crow with a problem that, to be successful, would take eight steps using tools in the proper sequence[127]. The crow had used the individual pieces of the puzzle before but had never put them all together, or in the required order.

Dr. Alex Taylor

Here is how it worked. The crow, in a large enclosure, could see a piece of food that he wanted. But it was inside a long, clear, plastic box with an open end. He could see that the box was too long for him to reach inside with his beak and get to the food.

New Caledonian Crow Using A Tool

Now, the crow knew about using a stick to extend his reach. Crows in New Zealand do this in the wild when they want to get an insect hidden in the bark of a tree. So, elsewhere on the test table, the crow could see a stick that was long enough to reach in and get the food, but it was also out of reach, in another, clear plastic box. So, how was he to get that box open?

In other experiments the crow, named 007, had managed to open that type of box by dropping a stone through a tube on the top side of the box. The weight of the stone was enough to push the downward hanging door open. On the table, he could see three wooden cages with stones inside. But again, he could not reach through the bars of the cages with his beak to get to the stones. He needed a small stick.

There was a small stick available elsewhere in the large cage. It was hanging, tied with a string to a perch. The crow flew to the perch and proceeded to pull up the string. He then pulled out the small stick (step1). He took the small stick and went to each of the three wooden cages with the stones inside and used the small stick to reach between the bars and bring out each stone (steps 2,3 and 4).

After a short time for consideration, he took the first stone and dropped it into the tube on the top of the clear plastic box with the large stick inside (step 5). It did not open the door as he expected.

He repeated that step with the second stone (step 6) into the same tube and box. Still nothing happened. Once more, he dropped in the third stone (step 7), and this time the weight of the three stones together was enough to open the downward hung door.

Now, 007 was able to reach inside and grab the longer stick. He walked over to the long, plastic box with the food inside and used the long stick to force out the tasty morsel (step 8). Success!

And it took *eight, sequential steps to achieve.* The researchers at Auckland claimed that this was *the first known demonstration of spontaneous, sequential tool use by an animal other than a human.* **And the creature who did it was a crow.**

One surprising capability demonstrated by wild crows is the ability to recognize and differentiate human faces. John Marzluff, a professor of wildlife science at the University of Washington, did some experiments that demonstrated this skill.[128] There was a flock of crows in a park near his campus. The research team caught and banded seven of the crows while wearing a rubber mask of an ugly, ominous- looking caveman with a large brow and protruding teeth.

For the next eight years, whenever someone would don that same mask and go to the park, those crows would spot the masked intruder and become agitated and begin scolding in alarm. Without the caveman mask, or even wearing a different control mask, the crows would not pay attention to those same people. Impressively, *the crows developed this behavior in just one trial.*

Professor John Marzluff

Even more interesting is that, somehow, *the crows communicated with each other so that, over time, most of the flock developed the ability to recognize the "cave man" whenever he would appear, even in a crowd. It is a good example of cultural learning in non-human animals, through socially transmitted behaviors.* Who would have guessed that crows were able to exchange such detailed information so well?

In another example of advanced intelligence, research scientists at the University of Tuebingen in Germany found that *crows can recognize the number of things in a set*[129]. They trained the crows to recognize groups of dots, from one to seven. Then they changed the sizes and arrangements of the dots, and the birds still recognized the number.

Neuroscientist, Dr. Helen Ditz, said, "When a crow sees three points, seeds or even hunters, single nerve cells recognize the 'threeness' of the objects." And this is exactly how primate brains perform, she said, which is interesting since the brain structures of the two species are so different.

Counting Crow

But it is not just Corvids in the Avian group that demonstrate unusual levels of intelligence compared to mammals. Many species of parrots also perform amazing feats of intellectual prowess.

The African Grey Parrot

In the section on the *Great Apes*, it was reported that Dr. Josep Call, of the Max Planck Institute for Evolutionary Biology, conducted a series of experiments with a group of Apes. Each animal was first presented with two cups. The animals could see that one cup was empty and one contained a favorite food.

Then they were taught *to identify the cup with the food by listening as the two cups were shaken. The cup with the food made a sound when it was shaken, the other was silent.*

In later experiments, the Apes demonstrated that they could *make* **an inference** *about which of the two cups contained the food* by listening to **just the empty cup being shaken**. Obviously, it made no sound. They understood that, therefore, *the food was in the other cup* and were rewarded when they selected it.

Other animals, such as monkeys and dogs, were given the same shaking task with logical inference, but none succeeded. *That is, until it was tried with African Grey parrots.*[130]

Dr. Christian Schlögl of the University of Göttingen in Germany led a similar, shaken cup study except using African Grey parrots as the subjects. Like the apes, the Greys had to make a two-step deduction. First, the presence and absence of noise had to be connected to the presence or absence of the reward, respectively.

This information then had to be used to deduce that the absence of noise in one container was predicting the presence of the reward in the other container. The Greys, for the most part, succeeded in this test *demonstrating that they were on par with apes in some cognitive abilities.*

Dr. Christian Schlögl

African grey parrots have become generally well-known as being extremely intelligent, thanks to one individual named Alex. The bird belonged to Dr. Irene Pepperberg who is now an animal cognition researcher with Harvard University.[131] She and Alex have been seen in numerous documentaries, television programs and presentations around the world demonstrating the parrot's broad range of capabilities. Dr. Pepperberg's widely read book, **Alex & Me**, made both of them household names.

Dr. Irene Pepperberg

Dr. Pepperberg received her Ph.D. in chemical physics from Harvard, but early in her career decided to switch her field to animal cognition studies.

In 1977, she purchased Alex from a pet store in Chicago and chose his name as an acronym for "Avian Language EXperiment".

Pepperberg began her studies with Alex at the University of Arizona and continued for over thirty years and three universities until Alex passed away unexpectedly in 2007. Since then, Dr. Pepperberg and her associates have continued the research with a number of other greys.

Alex, the African Grey Parrot

Over the course of his life, Alex:

- learned over 100 words

- recognized and could demonstrate that he knew seven colors, five shapes, and up to fifty different objects by name

- recognized quantities up to six

- understood the *concept of zero*

- understood the concepts of "bigger", "smaller", "same", and "different",

- learned, on his own, how to tell the researchers where he wanted to go.

- showed comprehension of personal pronouns; he used different language when referring to himself or others, indicating the concept of "I" and "you".[132]

Alex could identify a key as a key no matter what its size or color and could determine and communicate how a particular key was different from others. His accomplishments supported the idea that birds may be able to reason on a basic level and use words creatively.[133, 134]

Some academics, however, are skeptical of Pepperberg's findings, asserting that, without adequate data or peer-reviewed publications concerning his data, Alex's communications could be the result of operant conditioning.[135]

However, other studies, such as the shaken can test at the University of Göttingen, were peer-reviewed and do *provide support for the high level of intelligence displayed by African greys.*

Canines

Given the number of cortical neurons found in canines, including wolves and dogs, their potential for intelligent behavior seems to be small compared to some other mammals, like apes and Cetaceans.[XXXVIII] Yet, they are the creatures with whom we have the most up close and personal contact, at least in their domesticated versions. As a result, much work has been done in studying them. And they do not disappoint us with their abilities to perform. Some in rather amazing ways.

Wolves, for example, appear to have a relatively high level of intelligence. The Vanderbilt research team, headed by Suzana Herculano-Houzel, that counts cortical neurons in animals has not yet had the opportunity to count the neurons in a Wolf's cerebral cortex. But we do know that the wolf's brain is about 17% larger than that of a dog of comparable size. Since their brain structure is similar, we can assume that wolves have 17% more cortical neurons.   And, therefore, should have more potential for intelligence than a dog.

From observations in the wild, wolves appear to be curious about their environment and show a high degree of adaptability to varying conditions. They learn quickly and remember what they learned for long periods of time.

Professional studies of wolves' cognitive abilities have demonstrated that they can, indeed, perform at a higher level than dogs.

[XXXVIII] See the Introduction to this section. Dogs average about 525,000 cortical neurons. A German Shepherd size dog has 885,000, whereas Great Apes are in the 8 billion range.

Gray Wolf

For example, researchers at the Wolf Science Center of the Veterinary Medicine University in Vienna, Austria say that *wolves have a better understanding of causal actions than dogs.*[136] The researchers conducted the "shaking can" inference test[XXXIX] with a group of wolves and a group of dogs. Both Great Apes and Grey Parrots (see those sections) had previously succeeded in the experiment.

In this wolf and dog study, the animals had to make a choice between two objects, one of which contained hidden food while the other was empty. The wolves and dogs were both first taught that the object with the food inside made a noise when it was shaken, and that the empty one made none.

Once they learned that, they were shown both objects. The object that did *not* contain the food was shaken by the researchers and, of course, was silent. Then the animal had to choose which one contained the reward.

The wolves made the inference. They recognized that the silent can was empty, and even though they did not hear the rattle in the other can, they inferred that it did contain the food. So, they ignored the silent can and chose the other one that did contain the food. *The dogs, on the other hand, were not able to choose correctly.*

[XXXIX] See the sections above that discusses the Great Apes and Grey Parrots.

In many other cognitive tests, however, dogs have been shown to be surprisingly intelligent. *What has always astonished me was a dog's ability to acquire and understand language.*

For several years, I observed closely two dogs, both Weimaraners. I found that each had the ability to learn a surprising number of words, ideas, and concepts just by watching and listening. This is the same process used by a child to acquire language. For both of my dogs, I recorded the words and concepts they understood, and the number amounted to well over one hundred.

A dog who far exceeded the language skills of my dogs was named Chaser. She was a Border Collie owned by Dr. John Pilley, a Professor Emeritus of Psychology at Wofford College in South Carolina.[137]

Dr. John Pilley and Chaser

Chaser learned the names of objects through long, intensive training sessions. In total, the dog mastered the identity of more than *one thousand* proper nouns over her fifteen-year lifetime.

As she progressed, the training became easier when Chaser had an "a-ha moment". She realized that when Dr. Pilley said "this is," he was going to name something. She began learning names in one trial.

More than just words, she learned concepts including groups of categories. Pilley's daughter, Bianchi, said in an interview, "She (Chaser) had 30 balls and she knew all of them by a proper noun and also by category. You could ask her to find another ball, and she knew adjectives like bigger, smaller, faster, and slower."

Of course, anecdotal evidence, such as Chaser's history, is not scientific. But research scientists have reached the same conclusion, that dogs can easily learn object names and related concepts.

Another Border Collie, named Rico, was reported by his owners to understand more than two hundred simple words. A group of researchers headed by Juliane Kaminski of the Max Planck Institute for Evolutionary Anthropology in Leipzig, Germany studied Rico and wrote in **Science** that these claims were justified.[138]

During their tests, Rico retrieved an average of 37 out of 40 items correctly. In addition, he could remember items' names for four weeks after his last exposure. The researchers' bottom line conclusion was that Rico's vocabulary was broadly comparable to that of language-trained apes, dolphins, sea lions, and parrots.

Yet another group of researchers confirmed my observation that *dogs can learn language by watching and listening, much as does a human child.* Professors at the Department of Animal Science at De Montfort University in the U.K., Sue McKinley, and Robert Young conducted such a study. Their results were published in the journal, **Applied Animal Behaviour Science.**[139]

In this study, dogs *learned the label, or name, attached to a particular object only by listening.* The method used was *the Model-Rival approach*, the same technique used to teach language to Alex, the Grey Parrot.

In this approach, the dog watches its owner interacts with another human being. During the interaction, the owner and the "model-rival" discuss *the name of a particular object* that they want the dog to learn. During the discussion, the name of the object is mentioned several times. Then the owner throws the object and *commands the "model-rival"* to fetch it.

The dog has a strong interest in the interaction because it wants to know why its owner is interested in that person. In essence, the dog sees his "rival" for its master's attention as a model for a particular behavior that the researchers want the dog to learn, i.e. *the name of a particular object. And the reward for the dog is that it can model that behavior and then receive the owner's attention.*

After the training, the object is thrown by the owner *into a group of other, similar objects,* with the command to the dog to *fetch that particular object by its name.* The dog finds the correct object in the group and returns it to the owner.

Stanley Coren is a professor of psychology at the University of British Columbia. He wrote a book, **The Intelligence of Dogs** in 1994.[140] Coren used inputs from American and Canadian Kennel Club judges of obedience trials to rank the most intelligent dogs. He found that Border Collies, like Chaser, were consistently named *in the top ten by all of the respondents and were ranked the overall number one breed for intelligence.*

Stanley Coren

Border Collies were the subjects of a study to determine *if dogs had a g-factor similar to humans.*[141] [XL]. A large number of studies have demonstrated that in people, cognitive abilities overlap, yielding an underlying '*g*' factor which explains much of the variance.

Arden and Adams assessed individual differences in cognitive abilities between sixty-eight border collies. The objective was to determine if dogs had a general intelligence factor, "g", similar to that found in humans.

The focus of their experiments was on how well a dog's score *on one* of the three problems presented to the individual *predicted its scores on the other two tests.* It should be noted that *the other two tests involved quite different mental abilities.*[142]

[XL] See "How Intelligence is Measured" in this book.

Rosalind Arden

When the researchers did the statistical analysis, they found that dogs which were good on the first problem were also good on the other two classes of problems. While dogs who were not so good on the first problem did considerably less well on the other two tests.

Arden and Adams concluded that the results provide *strong evidence for the idea of a general factor of intelligence in dogs.* They found that smart dogs were generally proficient at everything, while not so smart dogs did generally poorly on most measures.

Cephalopods

Cephalopods, such as octopuses[XLI] and squid, have nervous systems that are radically different than that of vertebrates. Yet, they rival most vertebrates in their cognitive abilities and are *an important example of advanced cognitive evolution in animals.*

It has been several hundred million years since humans and cephalopods had a common ancestor. As a result, we have evolved very differently and are as unalike physically as one can imagine. Yet, *we humans and cephalopods share one especially important trait, a high degree of intelligence when compared to most other animals.*

The octopus is an invertebrate, meaning that they lack a spinal column, or, in their case, any bones at all. While the most common

---

[XLI] Octopuses and octopi have both been used as the plural of octopus, but "octopuses" has become the most accepted form in English.

octopuses may reach three feet in length, the Giant Pacific Octopus grows up to sixteen feet.

Giant Pacific Octopus

Octopuses are both predators and prey. They are *masters of camouflage* and can hide themselves from other predators by instantly changing their color *and texture* to match the environment.

They can accomplish this feat because their whole body is covered with *color changing cells* that the octopus can alter in fractions of a second. And it is not just the color, *but they can shift the pattern as well.*

Furthermore, their skin can be reformed to match the texture of an object in their environment, like a rock or seaweed. Their *camouflage skills, therefore, are unmatched in the animal world* and far exceed those of the chameleon.

Perhaps the most surprising physical differences of octopuses with mammals, is that the cephalopod *has three hearts, eight arms and nine brains.* Unfortunately for them, however, *their life spans are truly short. Usually it is only one to three years.*

Dr. Jennifer Mather, a comparative psychologist at the University of Lethbridge in Alberta, Canada has spent the last thirty-five years studying the intelligence of octopuses.[143] She is among a group of researchers who believe that these animals have a surprising amount of brainpower.

The focus of her work is to understand how octopuses' thinking processes evolved and compare that with the evolution of cognition in vertebrates, such as humans, primates, and corvids (crows and ravens). Despite the extensive differences in the evolutionary paths of vertebrates and invertebrates, *she has found that octopuses do indeed learn from their environment and then apply that knowledge in ways that are very much like the vertebrates.*

Dr. Jennifer Mather

Dr. Mather wrote in the *Canadian Journal of Zoology* that:

"A cephalopod (octopus) is first and foremost a learning animal, using the display system for deception, having spatial memory (knowing and understanding its environment), (having) personalities, and (engaging in) motor play. They represent an alternative model to the vertebrates for the evolution of complex brains and high intelligence, which has yet been only partly explored."

Aquariums around the world which have kept octopuses as part of their exhibits have reported many stories of apparent intelligent behavior displayed by these creatures.

They found that this complex predator can map its environment through observation, and then adopt a plan to achieve its objectives.

Decades ago, the Brighton Aquarium in England learned about this capability first-hand. They had an octopus in a display tank. The wise cephalopod was able to get out of its tank at night when no one was around, crawl across the floor to the next tank, get inside and eat a lumpfish. Then, the eight-armed predator escaped the lumpfish tank, wriggled back to its own tank where it was found the next morning looking completely innocent. It took more nights of similar behavior, and the loss of several lumpfish, before the aquarium staff figured out who was responsible.

Beyond the intelligence required, it may be surprising that an octopus could physically accomplish this feat. The reasons are due to its unique characteristics. It has eight, strong arms with suction cups that gives it extraordinary mobility, even on land. *Each arm has its own brain* and can function independently.

In addition, the cephalopod has a talent for squeezing itself through small openings, even though the holes may be much smaller than its body. It only needs an opening large enough to clear its beak and some cartilage in its head. The rest of their body can be compressed to squeeze through. This capability has earned the octopus the nickname, the "Houdini of the animal kingdom".

Finally, octopuses normally breathe with gills, but when they are out of water, they can absorb oxygen through their skin by passive diffusion. This allows them to operate out of the water for several minutes while they make their way across the floor of the laboratory. Or, between tidal pools as they do in the wild.

Many scientists from around the world, in addition to Dr. Mather, have become intrigued with octopuses' behavior and comparative intelligence. One of these is Dr. Graziano Florito at the Dohrn Zoological Institute in Naples, Italy[144].

He did an experiment to demonstrate *how octopuses learn through observation*. He and his team placed two octopuses in adjacent tanks, separated by clear plexiglass so they could see each other.

Each cephalopod had a clear plastic box in its tank with a delectable crab inside. The boxes had three openings with covers that required different types of motions to be released. One octopus had previously learned how to get into the crab container, the other had never done it before.

The name of the inexperienced octopus was Candide, as in the novel by Voltaire. The young cephalopod was just caught that morning by a local fisherman.

The octopus could see the crab in the box but did not appear to know what to do about it. The octopus in the *other tank*, however, had been in the laboratory for a few days and had figured out how to open the box.

This creature immediately went about his task, opened one of the covers, got into the container and devoured the crab.

Meanwhile, Candide, while carefully observing his neighbor through the clear plexiglas divider, became extremely excited. He apparently was watching and learning what he needed to know to be successful.

At this point, the experimenters removed the crab containers from the tanks, and placed a black plexiglass cover between the two tanks so the octopuses could not see each other. Then they put back the clear box with the crab into Candide's tank.

Now, the young octopus knew exactly what to do. He immediately went to the same opening of the three in the box that the other octopus had used. He grasped that cover and made the same motion he saw the other octopus make. He got into the container and had his meal.

The experimenters tried repeating this same experiment with Candide several times. But each time they rotated the box so that the correct opening was in a different position. Candide went immediately to that opening, used the same motion, and successfully acquired his meal. *This demonstrated to the experimenters that Candide had not just learned to imitate a movement but understood what had to be done.*

In other experiments, octopuses were able to learn to recognize symbols. They were taught that, of three doors, the one that had the symbol of a cross, wherever it was located, was the one that would lead to a tasty shrimp. With each trial, the symbols on the doors were rotated, but the cephalopod always went directly to the door with the cross, ignoring, the doors with the triangle or the circle

Like crows, octopuses have been observed by their aquarium keepers as being able to recognize faces of people they either like or dislike. Dr. Roland Anderson of the Seattle Aquarium confirmed this capability in an experiment he conducted in 2010. [145]

One researcher fed and played with their subjects, eight Giant Pacific Octopuses, while another scientist, dressed in the same uniform, scratched the same octopuses with an irritating object.

The octopuses quickly recognized the two staff members by their faces. The cephalopods easily approached the staff member with the food, looking for their treats. But, when they saw the researcher that had irritated them, they fled. To show their disdain for that person, the octopuses sometimes used their siphons to shoot water at them.

Giant Squid are another, exceptionally large cephalopod that are believed to be highly intelligent. However, none of these creatures have ever been caught and kept alive or bred in captivity. So, little is known about them including their exact level of intelligence. But some information is beginning to emerge.

As seen in the Figure below, a giant squid has a mantle or torso, eight arms, and two longer tentacles. These are the longest known tentacles of any cephalopod. The arms and tentacles account for much of the squid's great length.

Giant Squid also have three hearts, exceptionally large, primate-like, stereoscopic eyes, blue blood, a sophisticated nervous system and a complex brain. These characteristics suggest that they are capable of being among the smartest creatures on Earth.

Giant Squid

In a study reported in the journal **GigaScience**, researchers detail *the genome of a giant squid* for the first time.[146] Based on this work, we now know that the giant squid has about 2.7 billion DNA base pairs in its genome. For context, humans have about 3 billion base pairs, as do gorillas.

One of the most interesting findings reported by the researchers was the identification of one hundred genes that belong to the *protocadherin family*. These genes are not commonly found in invertebrates. *But they are another sign that giant squid have complex and highly evolved brains.*

> *"Protocadherins are thought to be important in wiring up a complicated brain correctly. They were thought to be a vertebrate innovation, so we were really surprised when we found more than 100 of them in* **the octopus'** *genome (in 2015). That seemed like a smoking gun to how you make a complicated brain. And we have found* **a similar expansion of protocadherins in the giant squid***, as well."*

Collective Animal Intelligence

Introduction

Collective intelligence is a mysterious force that enables some animals to display complex actions that are fascinating to observe. These include the collective behaviors seen in a swarm of bees, an army of ants, a school of fish or flock of birds.

Animals congregate in groups and act in concert for many reasons. The *collective intelligence of a flock of birds*, for example, helps to protect them from predators, saves energy when flying with "drafting"[XLII], keeps them on track when migrating, and allows them to share food discoveries.

The individuals in the group often have to make rapid decisions about where to move or what behavior to perform in uncertain and dangerous environments. These creatures have evolved the means to accomplish these tasks and do it amazingly well.

So how does this collective intelligence work? Groups are composed of individuals that differ with respect to their informational status. Each organism typically has only relatively local sensing ability of the other organisms around it.

And the individual members are usually not aware of the informational state of the others. This includes such items as whether the other member is knowledgeable about a pertinent resource, or of a particular threat.[147]

Yet, the individuals base their movement decisions on locally acquired cues such as *the positions, motions, or change in motions, of the other creatures around it.*[148]

[XLII] **Drafting, or slipstreaming**, is an aerodynamic technique where two birds, or other moving objects, are caused to align in a close group, reducing the overall effect of drag due to exploiting the lead object's slipstream.

In an attempt to explore the Darwinian fitness[XLIII] *of animal group decisions, numerous scientific investigations have been conducted.*

The goals were to understand how individual creatures with a low level of intelligence can display a relatively high level of collective wisdom. Through the latest research, we are just now beginning to recognize the relationship between individuals and group-level properties that lead to these displays.

Animals can be even more adroit than humans in using collective intelligence, while avoiding the maladaptive herding problem[XLIV] common in humans. Bees, for example, are well known for their ability to make accurate, collective decisions as they search for food or new nests, while *avoiding maladaptive herding.*

Ants

Ants are among our most common insects. There are more than 12,000 species spread around the world, found on every continent, except Antarctica. An individual ant is not very bright, but ants in a colony, operating as a collective, do remarkable things.

An analogous phenomenon is how a brain operates. A single neuron in the human brain can respond only to what the other neurons connected to it are doing. But when all 86 billion of them operate together, they can be the brain of a human being with the potential intelligence of an Albert Einstein.

In an ant colony, there is no supreme commander giving out orders. The only job of a queen ant, for example, is to be the reproducing female in that society. In fact, she is usually the mother of all the other individuals in the colony.

Instead, each ant decides on its own what to do next, and still the result is the amazing collective behaviors that we observe. For example, as individual foraging ants leave the nest in search of

---

[XLIII]*Darwinian Fitness*: the relative probability of survival and reproduction for a genotype A, compared to a genotype B.

[XLIV] *Maladaptive herding* - when *incorrect, destructive knowledge* goes viral, and is picked up through the social media and copied by many members of the human group. This results in erroneous, group decision-making as opposed to collective wisdom.

food, they walk in random paths, hoping to come across something to eat. The action is uncoordinated and chaotic. They know they are seeking a source of food, but they have no idea where to look. They continue in this way until they get tired and have to return to the nest. *Or they stumble across food.*

When an individual ant finds a food source, it takes a bit of the food back to the nest, leaving a trail of pheromones[XLV] behind them to mark the path to the food source. The ants in that area of the nest understand that the returning ant found food and has left a trail. *A group of them follow the pheromone trail back to the food source.*

Over time, and many trips to the source, the ants leave more and more pheromone trails. This action tends to organize their collection of the food, *optimizing the best and shortest path between the food and the nest.* They do not do it by thinking through and planning their actions as would humans, but by using the uncoordinated movements of the individual ants.

The chaotic, random foraging of individual ants is replaced eventually with organized precision. *Working as one, the ants create the sort of distribution network that a traffic engineer could only dream of.*[149]

Other types of ants display similarly remarkable behaviors. For example, there are over *two hundred species of army ants* found primarily in jungle and forest environments. They are composed of colonies with millions of members. One trait of army ants is that they have no permanent home. They march each night in search of a new foraging ground.

One interesting *group capability of army ants is the building of functional structures with their bodies.* These structures perform a variety of uses. One instance is a "living bridge" for the army to cross a gap in the colony's foraging trail.

[XLV] A **pheromone** is a secreted or excreted chemical factor that triggers a social response in other members of the colony. Other ants will sense the pheromone and follow it to the food source.

Army ants forming a living bridge

Researchers from the New Jersey Institute of Technology and Princeton University noted in their field experiments that the ants continuously modify their bridges, such that the structures lengthen, widen, and change position in response to traffic levels and environmental geometry.

A team, headed by Professor Chris Martin, did a detailed study of how this occurs. The results were amazing.

What the researchers found was that *the final position and structure of the bridge was modified in response to a group-level, cost–benefit tradeoff, without any individual ant having information on global benefits or costs nor the ability to use that information to do the analysis.*

The trade-off *benefit* was the increased efficiency of the use of the foraging trail which was balanced by the *cost* of removing workers from the foraging pool to form the structure.

To reach this conclusion the research team constructed a *computer-based model to determine the bridge location that maximized the foraging rate.* This result qualitatively matched the observed movement of the ant bridges.

So, the collective of ants, without the benefit of the information or the computational technology available to the researchers, made the same decision that the humans would have made.[150]

Fish

When fish form groups it is called "shoaling". In the biological definition, this means the fish are not synchronized, swimming somewhat independently, but in such a way that they stay connected, forming a social group. *Shoaling groups* can include the same species of fish or be composed of disparate sizes and species.

A Shoal of Yellow Snappers

But, when the group swims in a more coordinated fashion, and all in the same direction it is called "schooling". While fish are not typically considered 'intelligent,' *the schooling behavior they exhibit can be incredibly complex.*

A *school of fish are synchronized in their swimming.* They are usually of the same species and same age and size, so they look very much alike. The spacing between members is precise. The school makes complicated maneuvers giving the impression that it has a mind of its own.[151]

A School of Mackerel

Schools have properties that are possessed by the school but not the individual fish. These are called *"emergent" properties* and they give an evolutionary advantage to members of the school that are not available to non-members.

Some of these significant benefits include defense against predators through better predator detection, and by diluting the chance of individual capture. In addition, as a group, they enjoy enhanced foraging success, and higher success in finding a mate.

It is also likely that shoaling and schooling fish benefit through increased hydrodynamic efficiency. This means they save energy by swimming in the group since the other fish around them are moving some of the water for them. This action is like the "drafting" by a race car behind the competitor in front.

Shoaling fish tend to select groups with fish that look like themselves. This is due to what scientists call the "oddity" effect.

If any shoal member stands out in appearance it will be preferentially targeted by a predator. So, it is best for survival in the group to not be "odd", but to blend in with the other members. *The result is that shoals tend to be homogeneous.*

Improved ability to find food is certainly a fitness advantage. Swimming in groups enhances foraging success.[152] In a study by Pitcher, et al, the time it took for groups of minnows and goldfish to find a patch of food was quantified. The more fish in the group, the shorter amount of time to find the food. This was also found to be the case in studies of other species like Atlantic Blue Fin tuna.

The reason for this advantage is the presence of many more eyes searching for the food. But the "find" needs to be communicated. Fish in shoals "share" information by monitoring each other's behaviour closely. *Feeding behaviour in one fish quickly stimulates food-searching behaviour in others.*

Bees

Individual foraging bees find their food using both scent and sight. They then return to the hive and communicate the location of the new food source to the other bees. *How they accomplished this feat was a mystery until the early part of the twentieth century.*

It was the German Austrian, animal behaviorist, Karl von Frisch, who did the pioneering work in understanding the collective intelligence of honeybees. In his 1927 book **Aus dem Leben der Bienen** (*The Dancing Bees*), he described how foraging bees use a "waggle dance" for communicating to the others in their hive. *It provides information to them about the location of the food source and its quality.*

When a honeybee finds food, it uses two tools to understand and communicate the location to the others. First, a bee can see polarized light from the sun even when it is cloudy. So, it always knows the sun's relative location. The bee has the ability to remember the location of the food *relative to the position of the sun in the sky.*

Dr. Karl von Frisch

Second, the bee has a bio-clock that helps it track how far it has flown and how much the sun has moved during the trip. So, when it returns to the hive, she can convey the direction and distance to the food source using that information.

The first data that is conveyed by the foraging bee is the distance to the source. When the food is nearby, say less than one hundred yards, the bee performs a round dance, as shown in the figure below, by travelling in circles once to the left and once to the right.

If the food is further away than one hundred yards, the bee performs a *waggle dance*. This involves the shivering of the abdomen in side-side motion. During this dance, the bee forms a figure eight. *Additional directions are provided by the strength of the waggle, the number of times it is repeated, the direction of the dance, and the sounds made by the foraging bee.*

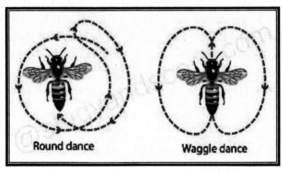

Honeybees' Dances

As a bee matures, her directions improve. It learns about the sun's changing path across the sky during different seasons of the year, and at different latitudes if her hive is moved.

The dance is also a type of rating system. The *quality level* of the particular source of food is conveyed to the others *based on the duration of the dance.* When a bee finds a good source of food, it dances for a long time. When it finds a relatively poor one, the duration of the dance is short or non-existent. *The longer the dance, the greater the number of bees that follow the directions to feed there.*

Unlike humans, bees avoid "maladaptive herding". If they are already using a particular location as a food source, but other foraging bees discover a much more productive site, the swarm will shift to that superior place. The bee's *collective flexibility*, therefore, is the key to avoiding the dysfunctional behavior of maladaptive herding.

Birds

Living along the coast of the Atlantic Ocean in Massachusetts, I see many species of sea birds. An interesting example is the cormorant. They are medium to large size, black, aquatic birds that make their livings by diving for fish. During the summer, I noticed that they remain as individuals or in small groups of two or three. They can often be seen standing on rocks in the harbor, drying their wings after long periods of diving.

But in the fall, when they migrate South, the cormorants gather in large groups of hundreds or even thousands. It is amazing to look up and see wave after wave of large cormorant flocks passing overhead. I have always wondered how they managed to find each other, organize into groups and, together, navigate to their destinations.

Cormorant drying out on a rock

Another bird that has fascinated me with their flocking behaviors is the starling. I recall driving along a straight, flat stretch of superhighway and seeing the swirling patterns of a huge flock of starlings (up to fifty thousand individuals) dancing across the sky about a mile in the distance. It took me a few minutes to understand what it was.

Scientists have also been intrigued by the starlings' performances and have conducted professional studies to better understand how they do it.[153] *They have found that by keeping track of their nearest neighbors and adjusting their movements accordingly, birds can respond rapidly and move with a high level of synchronization*

Swirling flock of Starlings

One such scientist is Andrea Cavagna, a statistical physicist at Italy's National Institute for the Physics of Condensed Matter (INFM). He said,

> *"It's a truly amazing sight. If you watch a flock of starlings under attack by a predator, the flock splits, then merges, and does all these incredible maneuvers to confuse the predator. How can they keep the flock's cohesion in the face of that strong perturbation—the attack?"*

Cavagna was part of a team of scientists and theoretical physicists in Rome who set up "StarFlag", a multidisciplinary, multinational collaboration to study the starlings' flocking behavior. The objective of the study was "to determine the fundamental laws of collective behavior and self-organization of animal aggregations in three dimensions," said Cavagna, the project's deputy coordinator.

> *"The project includes physicists and theoretical biologists who do computer modeling, biologists who study details of starling flight and behavior, and physicists and economists who work on extending the starlings' collective*

behavior patterns to such systems as cells in wound healing, aggregates of robots, and financial markets."

In Rome, the team made measurements of the starlings flocking behaviors using high-speed cameras. Then they conducted a computer analysis of the data to test how closely the actual flocking observations matched the simple rules of flocking developed by the computer scientists.[XLVI] Those rules are:

1. Separation – avoid crowding neighbors (short range repulsion)

2. Alignment – steer towards average heading of neighbors

3. Cohesion – steer towards average position of neighbors (long range attraction)

It was found that the Rules do indeed hold true in the case of starling flocking. But, Rule 3, the long-range attraction rule (cohesion), has some limitations.

It only applies to the nearest five to ten neighbors of the flocking bird, independent of the distance of these neighbors from the individual.

In addition, there is an anisotropy[XLVII] with regard to this cohesive tendency, with *more cohesion being exhibited towards neighbors to the sides of the bird,* rather than in front or behind. This is no doubt *due to the field of vision of the flying bird being directed to the sides rather than directly forward or backward.*

But what evolutionary benefit do starlings get from this flocking behavior? A study published in **Royal Society Open Science**[154] suggests that when starlings create a swirling mass, they are baffling their predators by using a phenomenon known as the "confusion effect".

[XLVI] Given the simple behaviour patterns of the *natural* individual birds, scientists felt they could be modelled by a relatively small set of mathematical 'rules. They hypothesized what those rules would be and built computer models to test them.

[XLVII] **Anisotropic** - having a physical property that has a different value when measured in different directions. A simple example is wood, which is stronger along the grain than across it.

When a raptor swoops in and tries to catch a single 'target starling' in a large, swirling mass of darting individuals, the predator becomes confused and, more often than not, misses their target.

Predator confusion is one of several functions of these large groups. In the winter months, huge flocks of birds gather to roost together, which provides protection and warmth from their collective body heat.

About 20% of all bird species are long-distance migrants. Migrating is a high stress, physically exerting activity. There is also a higher risk of predation on the journey. So, why do birds do it?

If no birds migrated, competition for adequate food during breeding seasons would be fierce and many birds would starve. Instead, birds have evolved different migration patterns, times, and routes to give themselves and their offspring the greatest chance of survival. *A significant, evolutionary benefit.*

So, a primary motivation for migration is improved access to food. For example, birds move South when their natural food sources become less available as the seasons change in the North.

There is a different, but related reason to migrate back North. It is the longer days of the northern summer. As the sun moves northward in the summer, the days are longer the further north you go. This is true up to the maximum amount at the north pole where there is twenty-four hours of sunlight.

This extra daylight provides an extended time for breeding birds to feed their young. And also, for them to produce larger clutches of eggs than those birds who remain in the tropics.

But exactly how do birds find their way from one precise location to the next? For example, I built my winter house in a pine tree forest in a rural part of Massachusetts. In the spring of the first year that we were there, a pair of Phoebes, a brown, medium size flycatcher, decided to build their brooding nest on a beam underneath our screened-in porch. In the fall, the pair left for their winter and migrated south as far as the Gulf Coast into Mexico.

Then, every year thereafter for the next forty years, in the first week in April, those Phoebes, or their descendants, came back and built another nest under the same porch. A very precise time and location after a trip of perhaps a few thousand miles.

So, on an early April morning, when we were having our first cup of coffee on the porch and heard the distinctive "phoebe" call, we took it as the first, true sign of Spring. The Phoebes had returned.

Many scientific studies have been done on a number of species to understand how birds do such feats. While researchers have found several different, interesting techniques that birds use for navigation, the details are still very much a mystery.[155]

The first possibility is the sensing of the Earth's magnetic field to identify direction. Many birds have unique chemicals or compounds in their brains, eyes, or bills that allows them to sense the Earth's magnetic field. Very much like a built-in compass. This helps them to orient themselves in a chosen direction for long journeys. But how exactly does that information enter and get processed in their bird brain?

Like human navigators, birds can also use celestial bodies to fly in a chosen direction. At night, star positions and the orientation of constellations can provide navigation information.

During the day, however, birds use the sun to orient themselves. When using the sun, it is necessary for the birds to compensate based on the time of day. So, they need an internal clock and the ability to do the computations.

But only sensing the direction would not be adequate to allow the birds to navigate to their destinations. Like humans, they also need some form of a chart, or map, to know which the correct direction is to select. Some scientists believe that the specific routes they follow along flyways may be genetically programmed[XLVIII], or learned to varying degrees.

[XLVIII] How are genetic "memories" passed on? A Tel Aviv University (TAU) discovered the mechanism that makes it possible. Their research was recently published in the journal **Cell**, and was led by Dr. Oded Rechavi from TAU's Faculty of Life Sciences and Sagol School of Neuroscience.

Birds appear to have the capability to map the geography of the areas over which they fly. Because birds follow the same migration routes from year to year, their keen eyesight allows them to map and remember their journey. Different landforms and geographic features such as rivers, coastlines, canyons, and mountain ranges can help keep birds heading in the right direction.

Some bird species, such as sandhill cranes and snow geese, learn migration routes from their parents and other adult birds in the flock. Once learned, in later years the younger birds can successfully travel the same route.

While some individual birds, such as corvids and parrots, can display significant intelligent behavior, *many more are able to employ collective intelligence to thrive in an unfriendly environment.* While these behaviors are of high interest to scientists, and many studies have been conducted, *there is still much to be discovered about how they do it.*

Animal Consciousness

Whether animals are "conscious" is still a matter of debate. We know that humans are conscious and that, somehow, it is associated with activity in the brain. But there is little agreement among scientists beyond that fact (see the section, "The Mystery of Consciousness").

In 2012, a group of neuroscientists attending a conference on "Consciousness in Human and non-Human Animals" at the University of Cambridge in the UK, signed the *Cambridge Declaration on Consciousness.*[156]

Cambridge Declaration on Consciousness

The absence of a neocortex does not appear to preclude an organism from experiencing affective[XLIX] states. Convergent evidence indicates that non-human animals have the neuroanatomical, neurochemical, and neurophysiological substrates of conscious states along with the capacity to exhibit intentional behaviors. Consequently, the weight of evidence indicates that humans are not unique in possessing the neurological substrates that generate consciousness. Non-human animals, including all mammals and birds, and many other creatures, including octopuses, also possess these neurological substrates."

So, it is clear that many neuroscientists believe that non-human animals can have a conscious life. The accompanying text to the Declaration added some other insights from these neuroscientists:

"The neural substrates of emotions do not appear to be confined to cortical structures. In fact, subcortical, Neural Networks aroused during affective states in humans are also critically important for generating emotional behaviors in animals. Artificial arousal of the same brain regions generates corresponding behavior and feeling states in both humans and non-human animals.

[XLIX] **Affect** refers to the outward expression of a person's internal emotions. For most people, there is congruence between affect and circumstance.

Wherever in the brain one evokes instinctual, emotional behaviors in non-human animals, many of the ensuing behaviors are consistent with experienced feeling states, including those internal states that are rewarding and punishing. Deep brain stimulation of these systems in humans can also generate similar affective states.

Furthermore, neural circuits supporting behavioral/electrophysiological states of attentiveness, sleep and decision making appear to have arisen in evolution as early as the invertebrate radiation, being evident in insects and cephalopod mollusks (e.g., octopus)."

I have owned dogs all of my life, so I have had a great deal of personal experience with them. Just as I know directly what consciousness is for myself, I know by watching my dog that he is also conscious.

I see it in his eyes when he is about to be fed his morning meal. He can smell the food being prepared for him and he starts to anticipate eating it. Philosophers call these qualia, or *individual instances of subjective, conscious experiences.* And the dog appears to go through them.

I have also seen him feel pain, another example of qualia. He runs a great deal among landscapes containing many trees and boulders and sometimes twists an ankle or pulls a muscle. As he is recovering over the next few days, he may squeal in pain when he moves the wrong way.

When the dog and I are relaxing, and he is curled up in his bed, he often falls asleep and dreams. I know he is dreaming by his actions. Most of the time, he is just completely relaxed. Occasionally, when he is, presumably, in REM sleep, he starts moving his legs as if he is running. And then he vocalizes a muffled growl or a bark as he "sees" prey or an intruder coming onto his territory.

Sleep is a natural state of being for humans and animals where consciousness and voluntary muscular activity are reduced. The function of dreaming in humans and dogs is to restore the brain-body system and ready it for the next day's activities.

Consequently, there is a special sleep state devoted to dreaming called REM (Rapid Eye Movement) sleep.[157]

When electrical activity recordings are made of the *dreaming, human brain* they *show that the area of our brain involved in sensory perceptions functions as it does when we are awake.*[158]

When we and dogs dream, we are virtually disconnected from the environment. We are not aware of what happens around us. Our muscles are largely paralyzed. Nevertheless, we are conscious, sometimes vividly so.

Another parallel of humans with animals' mental states is self-awareness. We know that animals, like Great Apes and Cetaceans, see themselves in mirrors and are thought by scientists to be "self-aware".[L] Most other animals, including dogs, do not pass the mirror test and are, therefore, thought to be *not self-aware.*

But this may be the result of prejudice, believing that all creatures are like us and, therefore, not using an appropriate method to test self-awareness for the particular species.

A problem with using the mirror test for dogs is that *dogs are more oriented toward their sense of smell than their vision.* A 2016 study claims that the *"sniff test of self-recognition"*[LI] supplies *significant evidence of self-awareness in dogs and wolves.*[159]

This study was done by a Russian team from Tomsk State University. The lead author of that study, Roberto Cazzolla Gatti, said:

> *"The innovative approach to test the self-awareness with a smell test highlights the need to shift the paradigm of the*

[L] **The mirror test** - is a measure of self-awareness developed by Gordon Gallup Jr in 1970. The test gauges self-awareness by determining whether an animal can recognize its own reflection in a mirror as an image of itself.

[LI] **"The Sniff-Test for Self-Recognition"** - the task involves measuring the time an animal spends investigating canisters containing their own urine versus those containing urine from another individual.

anthropocentric idea of consciousness to a species-specific perspective".

The researchers applied the sniff test to a group of four, captive, grey wolves living in male-female couples in two, separate enclosures located at *Wolf Park* in Battle Ground, Indiana, (Greater Lafayette), United States.

The study's authors recorded the increased time wolves spent investigating urine samples of *other wolves instead of their own.* The wolves were aware of their own urine and ignored it. *The results confirmed the evidence of self-awareness provided by other studies conducted with dogs.*

While my dogs also do not recognize themselves in a mirror, interestingly *they have recognized other dogs on television screens* and become extremely interested and involved in the action. They can watch for extended periods of time as long as dogs are visible and active. If they are sleeping a dog vocalization is heard from the TV, they jump to their feet and run to the television to see what it is.

This behavior was confirmed in a 2013 study published in the journal, **Animal Cognition.** It found that dogs could visually find images of other dogs among pictures of humans and other animals. They are also able to recognize on-screen animals and familiar sounds such as barking coming from the set. [160]

The bottom line is that many animals, from the Great Apes to our household pets to the squirrels and racoons in the back yard, are conscious. It is a mystery to scientists *how the brain creates* ***conscious experiences***, but we have them and so do our dogs and most other mammals. And most likely, do other intelligent creatures like octopuses.

Part III

Artificial Intelligence

III. Artificial Intelligence

Introduction

Our minds work in an extraordinarily, complex fashion. We use the information coming to us from our five senses, process all of that data in our brains, conclude what it means and then take the appropriate actions. This is done instantaneously, continuously, and unconsciously.

Driving a car is a good example of how well and quickly this process works with a human being in control.

When you drive your car at speeds up to 80 miles per hour, you are surrounded by other drivers, pedestrians, and four-legged creatures. Sometimes they seem to come at you from all directions. Weather conditions can range from a calm, sunny day with excellent visibility, to a blizzard with high winds and driving snow, so you can barely see the road.

To make the situation even more complex, you are travelling on many different types of roadways, from dirt roads to superhighways. There are road signs and traffic lights all along the way that must be read and obeyed.

Then there are the "rules of the road" that all vehicles must heed. But some do not, and this makes for many unpredictable situations to which the driver must adapt.

Sure, there are plenty of accidents, but considering the number of vehicles on the road at any one time, the quantity of mishaps is an exceedingly small fraction of the total.

Because of all of these factors, *it is hard to believe that a machine could ever drive a car as well as a competent human.* Yet, with the developments in Artificial Intelligence and sensor technology, we are fast approaching that reality. *It could become commonplace during this decade.*

In fact, the technology for autonomy is already here for driving cars and trucks on superhighways. *Industry predictions are that, with the further development of AI, sometime during the 2020s, all new cars will have the ability to drive themselves under all conditions.*

So, what is Artificial Intelligence (AI)? Basically**, *it is when a computer or a robot can autonomously perform tasks that are typically done by human beings and do them as well or better.***

Today's AI technologies include an array of software approaches that include Neural Networks, Machine Learning, Deep Learning, and Natural Language Processing. They allow AI systems to *learn* from examples and experiences, *recognize* objects and faces, *understand, and respond* to language, *solve complex problems, and make significant decisions.*

Yet *today* these systems are designed to *only work in narrow areas*, like helping physicians diagnose certain conditions, driving vehicles, allowing autonomous control of military drones, or providing part inspections in manufacturing operations.

A human level of AI will appear when the system has a general ability to learn, behave and work like a human being in a wide range of activities. This will require a much higher level of AI than has yet to be achieved. I*t is, however, close.*

Such a program, or even a humanoid robot, might be perceived by us to be conscious and even have emotions similar to ours, *whether or not they really experience those things.* In fact, it would have the Theory of Mind[LII] and would otherwise be indistinguishable from a human being.

A current gross classification of AI applications by scientists is Weak or Strong AI. *All AI applications today are considered to be "Weak",* or Narrow, although there are several levels of sophistication that are found in the current programs.

[LII] **Theory of mind** is the understanding that other people have beliefs, desires, intentions, and perspectives that are different from one's own. Possessing a functional Theory of Mind is considered crucial for success in everyday human social interactions and is used when analyzing, judging, and inferring the actions of others.

The *overarching goal of AI research is to create "Strong" AI applications* where the AI entity thinks and acts like a human in a wide range of situations.

Perhaps a slightly better classification system consists of three levels of AI.

- *Artificial Narrow Intelligence (ANI)* refers to a computer's ability to *perform a single task extremely well,* such as crawling a webpage or playing chess or driving a car. This is characterized as "Weak AI".

- *Artificial General Intelligence (AGI)* is when a computer program can perform *any intellectual task that a human can. Interacting with such a computer would be indistinguishable from an encounter with another human.*

- *Artificial Super Intelligence (ASI)* refers to *an AI that far surpasses human intellect by many orders of magnitude. Scientists are predicting that AI will reach this level sometime in the next ten to fifty years.*

  o *That point in time is called the Singularity.*[LIII] *Many people fear that the Superintelligence will be extremely dangerous and threaten the existence of the human race.*

  o We will look at this problem in detail later in this section.

All of the AI applications today, and as described in detail later in this chapter, are considered to be "narrow" or ANI. Many are very sophisticated and think and act like a human, but in a very narrow area of application.

[LIII] **The Singularity** is the point in time when machines become Superintelligent. This is when a technologically created, cognitive capacity is far beyond that which is possible for humans. At this point in time, *technology will advance beyond our ability to foresee or control its outcomes,* and the world will be transformed beyond recognition by the application of superintelligence to humans and/or human problems, including poverty, disease, and mortality. *The dangers are also unforeseen and uncontrollable.*

Many scientists are striving to reach the level of AGI, and some believe it will arrive during the 2020s. This stage of AI will emulate the abilities of human intelligence and be able to display its intelligence in a wide area of applications. *It will easily pass the Turing Test.*[LIV]

Artificial Superintelligence (ASI) is the Holy Grail of AI scientists. When we reach that level, sometime during this century, we will experience *the Singularity.* Beyond that point, *when the ASI is able to develop its own next generations, there will be an explosion of intelligence*[LV] that will be rapid and uncontrollable by humans.[161] **We will need to have safeguards in place prior to that time, or the ASI could, deliberately or not, end our existence.**

[LIV] **Turing Test** - a test proposed by Alan Turing. A computer is considered to display a human level of intelligence if the human conducting the test is unable to distinguish the machine from another human being by using the replies to questions put to both the fellow human and the computer.

[LV] **"Intelligence explosion"** - I. J. Good, an associate of Alan Turing, proposed a thesis. It states that a sufficiently advanced machine intelligence could build a smarter version of itself, which could in turn build an even smarter version, and that this process could continue to the point of vastly exceeding human intelligence.

The History of AI

The father of the digital computer, in the opinion of many scientists, is Charles Babbage (1791–1871). He was an English polymath[LVI] who was a mathematician, philosopher, inventor, and mechanical engineer. A contemporary term for such a person might by "Renaissance Man".

Babbage originated the concept of a digital, programmable computer in about 1834.[162]

Charles Babbage

His first entry into computing, however, was a mechanical computer that he called a *Difference Engine*. His idea for this device originated when he was asked by the British Astronomical Society in 1821 to improve the accuracy of the Nautical Almanac that was used for navigation applications. The objective of his work was to recalculate some of the tables of data found in the Almanac. As the job progressed Babbage identified a number of discrepancies in the calculations as they were currently generated. *He saw the need to have a more efficient and accurate way to do the computations and conceived an idea to do it mechanically.*

His concept, the *Difference Engine* , is *an automatic, mechanical calculator*. The name was derived from its use of the *"method*

[LVI] A **polymath** is a person who has a broad range of expertise in many areas. It is generally someone who has learned to think critically, sees the world through curious eyes and can make significant contributions to multiple fields.

of divided differences". This is a mathematical algorithm used to interpolate functions. As it turns out, some of the most common logarithmic and trigonometric functions used in engineering, science and navigation can be computed with the divided differences method.

Pictured below is a working model of the Difference Engine that was constructed in 1991 by the *London Science Mu*seum. It was the first one ever built based on Babbage's design and worked exactly as he had planned.

Difference Engine at London Science Museum

Unfortunately, Babbage was never able to actually complete the construction of his machine. This was the result of conflicts with his chief engineer, Joseph Clement. The dispute slowed the development of the project, and the British government finally withdrew its funding of the Engine because of the lack of progress.[163]

Several years later, in 1837, Babbage went on to design a *more advanced, general-purpose computer* which he called *the Analytical Engine.* He described it as the successor to the Difference Engine.

The logical structure of the Analytical Engine is essentially the same as that which has dominated computer design in the electronic era. *So, it was quite an achievement for Babbage,* working alone in the middle of the 19th century. *Yet, the Analytical*

Engine too was never built by Babbage, for reasons similar to the Difference Engine's lack of completion.

The beginning of the modern computing era was, arguably, in 1935. Alan Turing, a British mathematician, *created a theoretical model for computing machines* that could carry out calculations from inputs by manipulating abstract symbols on a tape. In later years, *this invention would inspire scientists to begin discussing and working on the digital computers that evolved into the machines we see today.*

Dr. Alan M. Turing

Turing, in his paper, described an abstract computing machine consisting of a limitless memory and a scanner that moves back and forth through the memory, symbol by symbol, reading what it finds and writing further symbols.

The actions of the scanner are dictated by a program of instructions that is also stored in the memory in the form of symbols. This is Turing's *"stored-program concept"*, and implicit in it is the possibility of the machine operating on, and thereby modifying or improving, its own program. *This process of self-improvement serves as the basis of AI.*

Turing's computing machine of 1935 is now known simply as the *universal Turing machine. All modern computers are in essence universal Turing machines.* As a result, *Turing is widely considered to be the father of theoretical computer science and artificial intelligence.*

Cracking the Enigma Code

During World War II, Turing played a major role for the Allies in cracking Germany's coded military communications. The Nazi forces employed an ingenious machine that was used to encode all their military command and control messages. The machine, called Enigma, had 159 quintillion (159 with eighteen zeros) possible settings. And the code was changed every twenty-four hours. The Nazis were confident, therefore, that their encryptions were unbreakable. Most cryptologists, at the time, agreed. *They were all wrong.*

The Nazi Enigma Machine

Early in the war, a group of British mathematicians, based in Bletchley Park, about fifty miles outside London, under great pressure by the Government to solve the problem, were successful in breaking the codes. A 2014 movie, called "The Imitation Game",[LVII] provided a highly fictionalized version of how it happened.

[LVII] "The Imitation Game" is the name of a game devised by Turing to determine if a computer has human level intelligence, now called the Turing Test. It had nothing to do with the content of the 2014 movie.

In the movie, Alan Turing, was shown as the primary genius behind the project. In real life, he played a major role, but it was a team effort and there were other mathematical geniuses who also made major contributions.

One of these was Gordon Welchman.[164] Like Turing, he was also a University of Cambridge mathematician and, in 1939, was among the first four recruits to work at Bletchley Park.

The cracking of the Enigma codes began in Poland several years earlier. The Cipher Bureau of the Polish Government was led by Dr. Marian Rejewski, also a mathematician.

The Poles developed an early form of a computer, which they called a *Bomba*, named after an ice cream dessert, to help break the Enigma codes. When the Nazis invaded their country, the Poles could not continue with their code breaking work, so they got their information out to England. It ended up at Bletchley Park.

Alan Turing started with the Polish design and, developed his first computer that *he called a Bombe*, a variation of the name of the Polish machine. This computing engine, completed in March 1940, shortened the time to break the daily code, but it was still not fast enough. *It had to reach a solution in much less than twenty-four hours because that is when the Nazis changed their Enigma settings.*

It was Gordon Welchman who came up with a technical solution to make the Bombe operate considerably faster. He added what was called a Diagonal Board to the design of Turing's rudimentary computer that dramatically speeded up the machine's processing time. Now, the Bletchley team could find the key to the daily Enigma code settings in an hour or two. *Then they could program their own, captured Enigma machines and read that day's Nazi messages.*

Turing and Welchman were the key figures at Bletchley Park, but there were literally thousands of people involved in the effort, and many of them were women. Most of these women were members of the Women's Royal Naval Service (WRNS). The British called them Wrens.

Eventually, there were hundreds of the top secret, Turing Bombes built by The British Tabulating Machine Company. Many were placed at Bletchley Park; others were set in various locations around Bletchley in order to distribute the risk of losing all of the machines in an air raid. The Bombes were operated around the clock by teams of Wrens.

Each day at midnight, the code breakers at Bletchley would identify "cribs", or best guesses of encoded words, from that day's first messages. For example, many messages ended with the words, "Heil Hitler".

The code breakers gave these cribs, identifying several letters in the daily code, to the Wrens who operated the Bombes. The young women then programed the Bombes using this information. One Wren would set the drums on the front of the Bombe, while the other would plug the wires in the rear of the machine to their appropriate settings.

If the crib information were correct, the machine would eventually stop running and the Wrens could read the Enigma machine settings for that day. With that information, all of the messages could be read. So, the Wrens played a key role in the work at Bletchley Park.

Welchman, while at Bletchley, was also credited with another major advance in cryptoanalysis. He called it *"traffic analysis"*.

In his approach, Welchman collected data-on-the-data (metadata) that was sent in the Nazi messages. He then used sophisticated, mathematical analyses to draw conclusions about the enemy's activities. It was an especially useful approach and is still employed by the United States' *National Security Agency (NSA)* as well as the British equivalent, the *Government Communications Headquarters (GCHQ)*.

Turing and Welchman were credited by Winston Churchill with having shortened World War II by at least two years. In the 1944 King's Birthday Honors list, Welchman was awarded the Most Excellent Order of the British Empire (OBE). Of course, his work was so secret at the time that it wasn't revealed why he had obtained the high honor. It was not until 1946, however, before Turing received a similar award.

Turing's Pioneering Years

After the war, Turing looked for opportunities to continue his work on computers. In 1946, he joined *Britain's National Physical Laboratory (NPL).* This is the national measurement standards laboratory of the United Kingdom. It is based at Bushy Park in London, England.

Turing used his computer design concepts to build the *Automatic Computing Engine. This was an early, electronic computer and one of the first designs to employ Turing's idea of a stored program.* Interestingly, the use of the word "Engine" in the name of the device was in homage to Charles Babbage and his computer designs, the Difference Engine, and the Analytical Engine.

In 1947, Turing gave a speech in London.[165] *It may have been the earliest public lecture to mention computer intelligence.* He said, *"What we want is a machine that can learn from experience".* He added that the *"possibility of letting the machine alter its own instructions provides the mechanism for this". This is a good description of today's AI.*

In 1948, Turing moved to the *Computing Machine Laboratory,* at the University of Manchester where he continued his computer development efforts along with the existing team of researchers.

The Laboratory, under chief designers, Frederic Williams and Tom Kilburn, was working on a device called the *Williams-Kilburn tube.* It was an early form of an improved computer memory based on standard, cathode ray tube technology.

In June 1948,[166] Turing, and the others at the lab, produced a small, prototype computer they called the *Manchester Baby.* It was officially known as the Small-Scale, Experimental Machine (SSEM). It too used Turing's concept of a stored program to operate the computer.

This development was quickly followed by the Manchester Mark 1, a more sophisticated version of the Baby. This machine's successful operation was widely reported in the British press, which used the phrase *"electronic brain"* in describing it to their readers.

Turing's achievement attracted the attention of the U.K. government which wanted to commercially develop the computer. They established a contract with the British electrical engineering firm of Ferranti International plc to produce a commercial version of the Mark 1. The resulting machine, *the Ferranti Mark 1, was the world's first, commercially available, general-purpose computer.*[167]

In 1948, Turing wrote (but did not publish) a report entitled "Intelligent Machinery". It is now available online.[168] *It is a detailed discussion of many of the key concepts of AI. One of these was the idea of "training" a network of artificial neurons to perform specific tasks.*

In 1950, Turing introduced a notion that is still discussed to this day. It came to be widely known as, *the Turing Test.* It is a check to determine if a computer has intelligence equivalent to that of a human being.

In his seminal paper, "Computing Machinery and Intelligence" he considers the topic of artificial intelligence.[169] The paper, published in 1950 in **Mind**, was the first to introduce his test concept. *He called it "The Imitation Game".*

Three participants are involved in the game, the computer, a human interrogator, and a human "foil". The job of the human interrogator is to determine, by asking questions of the other two participants, which one is the computer. All communication is via keyboard and screen, so no visual cues are available to the interrogator.

The interrogator may ask any question desired, and the computer is permitted to do everything possible to force a wrong identification. For example, if the computer is asked, "Are you a computer?" it can answer "No." Or it might purposefully provide an incorrect answer to a simple calculation in order to try to fool the interrogator.

The job of the foil is to help the interrogator to make a correct identification. During a test run, a number of different people play the roles of interrogator and foil. *If sufficiently many interrogators are unable to distinguish the computer from the human being,* then, according to proponents of the test, *it is to be concluded that the computer is an intelligent, thinking entity.*

In 1991, the New York businessman, Hugh Loebner, started the annual Loebner Prize competition, offering $100,000 for the first computer program to pass the Turing test. No AI program has, however, even come close.

Alan Turing died at the young age of forty-two under rather unfortunate circumstances. In 1952, he was prosecuted and found guilty of committing same-sex acts. There was an 1885 law, still in effect in the UK at that time, that "gross indecency" was a criminal offence. Turing accepted chemical castration treatment as an alternative to prison.

He died in 1954 from cyanide poisoning. An inquest determined his death to be a suicide, but it was noted that the evidence was also consistent with accidental poisoning.

In 2009, following an Internet campaign, British Prime Minister, Gordon Brown, made an official public apology to Turing on behalf of the British government for "the appalling way he was treated". A few years later, Queen Elizabeth II granted Turing a posthumous pardon.

The "Alan Turing law" is now an informal term for a 2017 law in the United Kingdom that retroactively pardoned men convicted under historical legislation that outlawed homosexual acts.[LVIII]

Turing has an extensive legacy of honors for his important achievements in computer science and AI. There are statues of him in various locations, and an annual award is named after him for computer science innovations. Most recently the British government honored him by placing him on the Bank of England £50 note, released in June 2021.

Image of the Turing Fifty Pound Note

---

[34] "Alan Turing law': Thousands of gay men to be pardoned". BBC News. 20 October 2016.

The Dartmouth 1956 Workshop

The Dartmouth Summer Research Project on Artificial Intelligence was a 1956 workshop widely considered[170] to be the founding event of artificial intelligence as a field of study. *It was the moment that AI gained its name and its mission. Out of this conference, emerged the initial players in the field.*

The idea for the conference was originated in the fall of 1955 by John McCarthy, a young Assistant Professor of Mathematics at Dartmouth College in Hanover, New Hampshire. His goal was to organize a group to clarify and develop ideas about "thinking machines." *He picked the name "Artificial Intelligence" for the new field.*

The conference needed funding, so McCarthy and three other prospective participants approached the Rockefeller Foundation. They wrote a proposal that states:

> *"We propose that a 2-month, 10-man study of **artificial intelligence** be carried out during the summer of 1956 at Dartmouth College in Hanover, New Hampshire. The study is to proceed on the basis of the conjecture that every aspect of learning or any other feature of intelligence can in principle be so precisely described that a machine can be made to simulate it. An attempt will be made to find how to make machines use language, form abstractions and concepts, solve the kinds of problems now reserved for humans, **and improve themselves**. We think that a significant advance can be made in one or more of these problems if a carefully selected group of scientists work on it together for a summer."*

The proposal itself is credited with introducing the term "artificial intelligence" into formal usage. The topics mentioned to be covered in the conference included computers, natural language processing, Neural Networks, and the theory of computation. These are still active topics in the AI field today.

Two of the other authors of the document were *Marvin Minsky and Claude Shannon, both MIT professors,* but Shannon was then working at Bell Labs. They were both well-known authorities in their fields and highly respected.

The funding was approved, and the conference was scheduled. The meetings were held that summer in a large classroom of the Dartmouth Math Department. The project lasted about seven weeks and was essentially an extended brainstorming session.

Those who attended would become the leaders of AI research for the next several decades. The field of AI was finally formed as an academic discipline.

Some of the Attendees including Marvin Minsky and Claude Shannon (Front left and right)

Many of the attendees of the 1956 conference predicted that a machine as intelligent as a human being would exist in no more than a generation, say about 1980. They were given millions of dollars by corporations and the Federal Government to make this vision come true. Here are some of their quotes:

- 1958, *H. A. Simon and Allen Newell*: "within ten years a digital computer will be the world's chess champion" and "within ten years a digital computer will discover and prove an important new mathematical theorem."[171]

- 1965, *H. A. Simon*: "machines will be capable, within twenty years of doing any work a man can do."[172]

- 1967, *Marvin Minsky:* "Within a generation ... the problem of creating 'artificial intelligence' will substantially be solved."[173]

- 1970, *Marvin Minsky* (in Life Magazine): "In from three to eight years we will have a machine with the general intelligence of an average human being."[174]

But they and their funders were to be sorely disappointed.

Eventually, it became obvious that the scientists at the conference had grossly underestimated the difficulty of developing an AI system with the intelligence of a human being. *By 1973, it was clear that there was an exceptionally long way to go to achieve what we now call "strong" AI.*

Artificial Intelligence: Boom and Bust

The US Congress was backing much of AI research through the Advanced Research Projects Agency. This agency was later renamed Defense Advanced Research Projects Agency or DARPA and still continued to fund advanced AI projects.

The government administrators, however, became disillusioned in the early seventies with the lack of progress in AI. They stopped the funding of *undirected* AI research. *This led to the first "AI winter" of research that lasted from 1974 until 1980 when another miniboom in AI occurred.*

The advent of a type of AI program called an "Expert System" changed the landscape. Expert Systems are programs that answer questions or solve problems about a specific domain of knowledge, *using logical rules that are derived from the knowledge of experts. These were the first AI programs that were perceived to have significant value by the users.* [175]

Knowledge based systems and knowledge engineering became a major focus of AI research in the 1980s. These programs will be covered in detail later in this Part.

The interest of Corporations and the U.S. government in AI again rose with these developments. But it soon became clear that the promise of these applications did not provide the value to the users that was originally anticipated. The result was that *the perceived worth of AI applications again declined in the late 80's, and the second AI Winter set in from 1987 to 1993.*

Something had to change to get private and government entities to again become interested in AI. And it did.

The mid-90's saw the advent of new, stronger software programs focused on narrow AI applications. These algorithms were coupled with the use of more powerful, higher speed computers, possessing vast amounts of memory, and utilizing immense data sets.

The combination allowed scientists to start to make some headway in advanced AI systems and applications. This was attractive to many users and research funding again started to flow.

Inside the AI community, however, there were significant disputes. The experts could not agree on the reasons for AI's failure to fulfill the vision and 1960's dream of human level intelligence. The arena became *highly fragmented* as competing subfields focused on particular problems or approaches.

For example, algorithms that were developed by AI researchers were used to solve problems in data mining, industrial robots, speech recognition, medical diagnostics, and even Google's search engine.[176] Some of these applications will be covered in more detail later in this Part.

Occasionally, AI researchers used different names that disguised the tarnished pedigree of "artificial intelligence".[177] Examples were *informatics, knowledge-based systems, cognitive systems, or computational intelligence.* Nevertheless, *AI was both more cautious and more successful than it had ever been before.*

Nick Bostrom, an Oxford philosopher, explains "A lot of cutting-edge AI has filtered into general applications, often without being called AI, because *"once something becomes useful enough and common enough it 's no longer labeled AI."*[178]

In 2010, a sub-field of AI research emerged, bringing back the original target of developing AI with human level intelligence. This time around it is called *Artificial General Intelligence (AGI).*

There are now academic conferences, university courses and new companies springing up to develop AGI. So, we have come full circle in terms of AI expectations. The ultimate goal is now *Strong AI.* This is how academics classify AI with consciousness and superintelligence. Think HAL 9000 from **2001: A Space Odyssey.**

Will the scientists be successful this time, or will it lead to a third AI winter?

AI Applications in the Modern Enterprise

Introduction

The digital transformation of many industries is well underway.[LIX] Over the past five years, the International Data Corporation (IDC), a major market research firm, has been documenting the rise of the digital economy.

This transformation is the result of competitive pressures. To maintain a meaningful competitive position, companies have to bring into their operations the capabilities provided by modern, digital technologies.

By 2023, IDC predicts that the global economy will finally reach "digital supremacy" with more than half of all GDP, worldwide, driven by products and services from digitally transformed enterprises.[179]

Now, industry is going through another, and even more powerful, transformation. **It is adopting a wide range of AI applications.** With digital processes in place, the next step is to improve the intelligence of those processes. This increases the level of automation as well as the efficiency and effectiveness of those processes.

AI transformation touches all aspects of the modern enterprise including both commercial and operational activities. Tech giants are integrating AI into their businesses.[180] For example, Google is calling itself an "AI-first" organization.

In addition to the tech giants, International Data Corporation (IDC) *estimates that a majority of other organizations will insert AI technology into their processes and products by 2025.*

Similar to the digital transformation, making the AI transformation will provide a major competitive advantage to the adopters. On the other side of the coin, companies who lag behind in bringing on AI will lose market share and profitability.

[LIX] **Digital transformation** is the process of using digital technologies, such as electronic tools, systems, devices that generate store or process data, to create new, or optimize existing, business processes to meet changing market requirements.

There are numerous applications in today's businesses of what is classified as Artificial *Narrow* Intelligence (ANI). We will cover a few of them to give you an idea of what AI is doing in the world today.

Expert Systems

An Expert System (ES) is a computer-based application that emulates the decision-making ability of human specialists. It is one of the oldest AI applications and led to one of the booms in AI funding back in the 1980s. The first ES was developed in 1965 by Edward Feigenbaum and Joshua Lederberg of Stanford University. *Their system was designed to analyze chemical compounds.*

Expert systems now have commercial applications in fields as diverse as medical diagnosis, petroleum engineering, and financial investing.

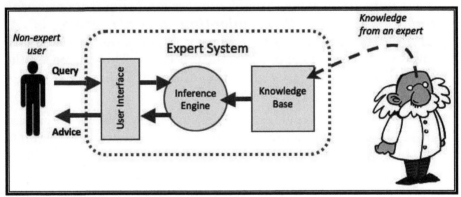

Elements of an Expert System

The internal structure of Expert System consists of three parts:

1. The **User Interface** – where the user inputs the data to be analyzed, typically using a keyboard. It also has the means for the output to be provided, usually displayed on a monitor and/or as a printout.

2. The **Inference Engine**- this is software that matches the users input with data contained in the knowledge base to reach appropriate conclusions. This is done using inference rules made of *If-Then statements.*

3. The **Knowledge Base** – contains data and facts in the specific knowledge domain, provided by the experts in the field.

A specific example in medical diagnosis is an Expert System developed at the *Learning Disabilities Clinic (LDC) of Boston Children's' Hospital* to *diagnose children with mathematics learning disabilities.* It is called the *Mathematics Diagnostic/Prescriptive Inventory (MDPI).* [181] [LX]

It is a good illustration of what can be done using the logic functions in the generally available Excel spreadsheet program based on a personal computer. The software is easy to work with and the computer set up is useful for managing the input/output.

A student is given an extensive, diagnostic examination consisting of a wide range of math problems, some using manipulative materials. There are up to 1,162 data points describing both skills' achievements and skills' approaches. The results of the exam are recorded during the administration, and the student's performance is then input into the system using the computer user interface.

The Knowledge Base of the Expert System incorporates the knowledge, research, and experience that the Learning Disabilities Clinic, Math Learning Specialists have acquired in over forty years in clinical practice, and decades of working in classrooms.

This Knowledge Base is further informed by consultations with other members of the Learning Disabilities Clinic team, including Reading and Language specialists, as well as neurologists, neuropsychologists, and psychologists.

Most Expert Systems use IF-THEN logic rules to represent the knowledge of the team, and this is the case with the MDPI

[LX] **Maria Marolda**, the head of the Math Team at the LDC, was both one of the experts and the programmer of the ES. She built the Inference Engine with the IF-THEN Logic functions found in the Excel spreadsheet program.

approach. Typically, Expert Systems can have a few hundred to a few thousand such rules.

IF, for example, the patient responds to a particular set of questions in a certain way, THEN they are deemed by the system to have a Concrete style of thinking, as opposed to Global or Linear styles. This determination, along with many others, has implications in the diagnosis, and then the remediation treatment of the patient.

Unlike Machine Learning, the Expert System *needs to be refined by the programmer* as it is used in practice. As actual results become available, the model's code is "tweaked" to better reflect the judgement of the experts in a wide range of situations. Over time, the ES is optimized.

With the MDPI, *the primary output of the analytic engine is the student's Mathematics Learning Profile (MLP)* which is provided on a separate sheet in the Excel workbook. It represents a portrait of a student's strengths and weaknesses in mathematics and provides a map for current and future instructional efforts.

By understanding the student's MLP, teachers can identify reasons why the student is doing poorly and why mathematics might be a struggle for them in terms of achieving grade appropriate skills. And then a suitable remediation program can be developed.

Manufacturing

Implementing AI in manufacturing facilities is increasing at a significant rate. According to Capgemini, a consulting and research firm,[182] more than half of Europe's manufacturers (51%) are implementing AI solutions, while Japan (30%) and the US (28%) are following in second and third places. Forced by competitive pressures, *the trend for AI adoption in manufacturing operations will continue and accelerate as the AI technology gets faster and cheaper.*

Intelligent Maintenance

Using *AI for intelligent maintenance of machinery* in manufacturing operations can provide quick returns for companies. The AI system, equipped with Deep Learning, is *trained* using data from historical machine failures. The AI then understands the elements that failed, and the reasons for those failures. Later, with appropriate sensor input, the AI is able to predict future malfunctions that can then be addressed early.

Remember in "2001: A Space Odyssey", Hal 9000 identified a module in the communications system that was about to fail and told the crew to replace it. That is exactly how these intelligent maintenance systems operate.

Sensors, at various points in the plant's equipment, continuously collect data on the operational factors that affect machine performance. This data is uploaded into the AI-based system.

The AI analyzes the data, predicts upcoming failures, and makes a variety of recommendations to the maintenance crew. One of the recommendations is an *optimal time* to conduct the maintenance in order to *minimize production downtime.*

Automated Product and Part Inspection

AI Machine Vision systems provide increased efficiency and effectiveness for *product inspection operations*. The elements and general process flow of such a Machine Vision system are shown in the Figure below.

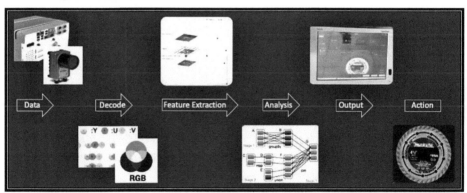

Typical Machine Vision System in Production Operation

A typical application might be for an assembly line on a production floor. After an operation is performed on a part, the camera of the vision system is triggered to capture and process the part's image using the computer vision software. The software may be trained to check the position of the part, its color, size, and shape. The

Machine Vision system can also look at, and decipher, a standard barcode, or even read printed characters.

After the part has been inspected and analyzed, a decision is made regarding the quality of the part, and a signal is generated to determine what to do with the analyzed part. If there are flaws, the part might be rejected into a container or an offshoot conveyor.

Alternatively, if the part passes the inspection, it moves on through more assembly operations. The part and its inspection results are tracked throughout the manufacturing process.

Other typical uses for Machine Vision systems in manufacturing include:

- Test and calibration,
- Real-time process control,
- Data collection,
- Sorting/counting.

Manufacturers use Machine Vision systems instead of human inspectors because the automation is better suited to repetitive inspection tasks. It is faster, more objective, and works continuously and tirelessly without the need for coffee breaks.

Industrial Robotics

Today, industrial robots are in wide usage in many industries. They have been performing difficult-for-human tasks, using *traditional* electronics capabilities, for decades. *But recently, the industry has begun to adopt AI which is promising to transform the application. For example, machine learning will allow the robots to learn their tasks faster and more efficiently than using traditional programming.*

The use of robots began in the 1960s.[183] Unimation was the world's first robotics company. It was founded in 1962 by Joseph F. Engelberger and George Devol and was located in Danbury, Connecticut. Their product was called *Unimate.*

My group at Arthur D. Little, Inc. worked with Mr. Engelberger to identify application opportunities for his robot. The first adopter was General Motors who used it in one of their plants in New Jersey for some minor assembly tasks.

ASEA, a Swedish firm, developed the ASEA IRB in 1975 that was the first fully, electrically driven robot. It was also the first microprocessor-controlled robot that applied Intel's initial chipset.

But major growth of industrial robotics did not come until after 1980. Yaskawa America Inc. introduced the Motorman ERC control system in 1988. It had the power to control up to 12 axes,[LXI] which was the highest number possible at the time. FANUC robotics from Japan also created the first prototype of an "intelligent" robot in 1992.

Industrial robots on automotive assembly lines are now large, powerful machines moving and operating on heavy pieces of the vehicle. The robots are bolted to the floor, and, for safety's sake, humans have to be kept away from the process, so the robots are fenced in.

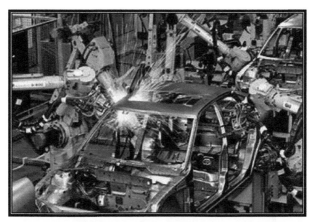

Assembly and welding robots in an auto factory

All of these industrial robots, until recently, *have been programmed to perform a single function composed of multiple steps.* This is accomplished by writing complex code, which takes a long time, and requires formidable programming expertise. *This makes it exceedingly difficult to adapt the robot to new situations.*

[LXI] To provide flexibility of motion, there are many points in an industrial robot, *called axes*, which can *move in different directions.* For example, the base can move from side to side or back and forth. Points on the lower and upper arms can move in multiple directions including back and forth and side to side, as well as having a rotating wrist action.

Collaborative robots (cobots) are relatively new and quite different. Although they too must be programmed, the process can be so simple that anyone can do it using a PC keyboard.

The *first cobot* was installed in 2008 at Linatex, a Danish supplier of a *wide range* of plastics and rubber products. The company needed an automated solution that was *flexible* enough to serve their versatile product line.

With the pioneering cobot technology of *Universal Robots, founded in Denmark in 2005,* Linatex management finally got what they wanted: a low-cost robot that can be easily re-programmed for new production batches.

Another advantage of the cobot is that it can be safely placed on the manufacturing floor, as opposed to being locked behind a fence.

Jørn Trustrup, Product Manager at Linatex said,

> *"Now, collaborative robots are a fast-growing trend in industry. The next step is the addition of artificial intelligence (AI) to increase their capabilities.* AI will expand the range of tasks the robots can perform and reduce the need for work-environment adaptations to accommodate them."

Cobot working with human partner

The robots use sensors and software for control and perception. This technology determines where and how the robot should move. With optical sensors for perception, the robot can understand and react to its surroundings. This includes keeping its human partners safe.

More specifically, data to operate the robot comes from a number of sensor technologies. They include integrated laser scanners, 3D cameras, accelerometers, gyroscopes, wheel encoders, and other such sensor technologies. The sensors provide the information necessary to produce the most efficient decisions for each situation.

The mobile versions of cobots are able to dynamically navigate around the shop floor using the most efficient routes. They also have environmental awareness so they can avoid obstacles or people in their path. Finally, they can automatically re-charge their batteries when needed.

AI for collaborative robots is focused primarily on Machine Learning (ML) and vision systems. These technologies are dramatically extending earlier, sensor-based capabilities. *ML progressively improves the cobots skills through experience gained over time.* The AI algorithms, through learning and data analysis, enable the cobots to make predictions and, thus, make their own decisions.[184]

Motion and manipulation of objects, *implemented through AI*, are basic skills for a cobot. They allow the cobot to handle objects with grippers at the end of their arms.

AI is essential for mobile cobots that need to perform navigation, localization, mapping, and planning. As with perception, these skills heavily rely on sensors for the input data, while the AI analyzes and directs the cobot.

Natural Language Processing is now the subject of intense development for cobots. The objective is to allow cobots to converse with and thus learn from their human partners. Much more work needs to be done in this area before the robot and human are able to communicate in a human-to-human manner. But significant progress is being made.

With increasing computing power becoming available, two other advanced AI capabilities are being adapted to cobotics to take it to the next level. They are *Neural Networks and Deep Learning.*

A *Neural Network* is designed to achieve advanced learning skills without the need for any type of programming. Effectively the system mimics many of the abilities of the human brain.

Deep Learning is the most advanced form of Machine Learning; like neural networks, it is concerned with algorithms inspired by the structure and function of the brain. Deep learning gets its name from its large number of layers and is basically a "deep" Neural Network.

Putting these key, advanced AI technologies to work in robotics will usher in the next paradigm in that field.

What most people think of when you say "robot" is a humanoid machine. The company with the most advanced mobile robots, including humanoids, is *Boston Dynamics*, an MIT spinoff. It started in 1992 with a group of entrepreneurs forming a start-up that was funded by venture capital. It has been sold and is now owned by Hyundai, the South Korean, automobile manufacturer.

Boston Dynamics (BD) is a world leader in *mobile* robot technologies. It specializes in high-performance robots equipped with perception, navigation, and some limited intelligence.

Their major commercial product at this time is *Spot*, an agile robot that navigates terrain with significant mobility, allowing the user to automate routine inspection tasks and data capture, safely, accurately, and frequently. It can even use its arm to open a door and then walk through it.

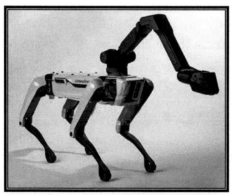

Boston Dynamics Robot 'Spot"

A much less glamourous new product *could become BD's most commercially successful market entry* and turn BD into a profitable company. It is called *Stretch. It has one job, to move boxes in an industrial/warehouse environment.*

Boston Dynamics *Stretch* Moving Boxes

As seen in the above Figure, Stretch has a box-like base with a set of wheels that can move in all directions. This means it has the flexibility of movement that it can be used in a number of applications in different locations in the factory or warehouse depending upon the needs.

On top of the base are a large robotic arm and a perception mast. The robotic arm has seven degrees of freedom and a suction pad array that can grab and lift boxes.

It uses AI capabilities with its computer vision system and, a form of Deep Learning to learn its specific tasks. The perception mast has the computer vision–powered cameras and sensors to analyze its surroundings.

Robotic Process Automation

Another application using AI in business operations is *Robotic Process Automation* (RPA), a *software technology* (no physical robots are involved) that *emulates the actions of a human while working on a computer.*

Reading and evaluating hundreds of resumes is one example of a task particularly suited to a software robot.

These software robots can understand the information on the screen, identify and extract data from that information, perform a wide range of defined actions, complete the right keystrokes to input new information or judgements, and then navigate to another case or item.

But software robots can do it faster and more consistently than people, without the need to get up and stretch, or take a coffee break.

Robotic Process Automation capabilities are helping organizations with a number of other diverse applications. They automatically process and store data without having to perform manual data entry, generate financial reports without spending considerable amounts of time in Excel, and execute customer outreach campaigns without spending hours in a Customer Relationship Management program.

Some of the currently *more advanced* RPA systems have adopted cognitive technologies such as Machine Learning, Natural Language Processing and speech recognition. Such skills provide them with additional intellectual capabilities that are characteristic of humans. These include the ability to learn, acquire knowledge or skills, communicate with humans through speech and then solve problems with that knowledge.

Collective Robotic Intelligence

Humans, and some animals such as *ants, bees, birds, and fish, display collective intelligence (*see Collective Human Intelligence in Part I, and Collective Animal Intelligence in Part II.) Collective intelligence is *group* intelligence that emerges from the collaboration and collective efforts of many individuals.

Nature has proven that when individual creatures with low intelligence levels, like ants, collaboratively work together as a unified system toward a common goal, they reach that goal faster and more accurately. In other words, they are smarter together than they are on their own.

In the case of robots, like their animal counterparts, *their relatively low, individual intelligence* is amplified when they *collaborate locally with each other, and with their surrounding environment, while adhering to a general set of algorithmic rules.*

The expression, *Swarm Intelligence (SI)*, was introduced by Gerardo Beni and Jing Wang in 1989.[185] *Swarm intelligence* refers to the *set of algorithms* used to direct the individual robots' activities. When the algorithms are used with robots it is called *swarm robotics*.

Like the animal models of collective intelligence that we considered in Part II, the robotic agents follow these quite simple rules. There is no leader directing the activities of the individuals. *Instead, local, and some random, interactions between such agents lead to the emergence of "intelligent" global behavior.*

The positive results are unknown to the individual agents but are achieved by the swarm behavior of all of the agents.[186]

Kilobots from Harvard University

Shown in the Figure above is a swarm of "Kilobots". The Kilobot is a 3.3 cm tall, low-cost, swarm robot developed by Radhika Nagpal and Michael Rubenstein at Harvard University.

Their purpose is for research in swarm robotics. The Kilobots can act in groups, up to a thousand, to execute commands programmed by users that could not be executed by individual robots.

A significant advantage of working with the Kilobots is their low cost. The total cost of parts for a single robot is under $15.

There are many potential *applications for Swarm Robotics.* They include uses in healthcare, military, and search and rescue missions. *The healthcare and military applications will be covered in those sections.*

One of the most promising uses for swarm robotics is for search and rescue (SaR). Swarms of robots of different sizes are sent to places that rescue-workers cannot reach safely. The robots have onboard sensor and processor systems used to devise complex search patterns to thoroughly explore the area.

One advantage of using a distributed system of robots for Search and Rescue is to increase the robustness of the swarm. There is less dependency on a particular part of the swarm. If one member of the swarm is damaged in the hostile environment, others can take over its task.

Another important feature of Search and Rescue robots is versatility. During a mission, an individual robot will face unpredictable situations and it must be able to manage those challenges.

One approach to deal with this problem is *self-reconfigurable robots. They provide* versatility and adaptation within the swarm. Self-configuration enables the robot to change its structure depending upon the specific characteristics of the environment.[187]

Self-reconfigurable robots are constructed of robotic modules that can be connected in many different ways. The modules are self-contained with their own microprocessor, sensors, and actuators. These modules move in relationship to each other, which allows the robot as a whole to change shape. This shapeshifting makes it possible for the robots to adapt and optimize their shapes for different tasks during the SaR mission.

Thus, a self-reconfigurable robot can first assume the shape of a *rolling track* to cover distance quickly, then the shape of a *snake* to explore a narrow space, and finally the shape of a *six-legged hexapod* to carry an artifact back to the starting point.

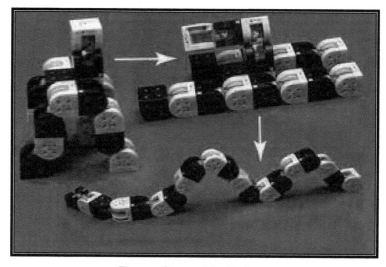

Reconfigurable Robot

Automotive

Tesla automobiles are among the most automated vehicles on the road today. Their *Autopilot* is an advanced driver assistance system designed to enhance the safety and convenience of the owner by reducing the driver's workload. Most other brands of luxury vehicles are now equipped with a similar system.

The Tesla, however, is equipped with a larger range of sensors than most other brands and has an optional package for "Full Self-Driving Capability". The sensors include eight external cameras, a radar, and twelve ultrasonic sensors. They are all coupled to a powerful, onboard computer.

In any mode, the vehicle is *not autonomous, and the driver must be fully alert with hands on the wheel in order to take over at any time.*

Tesla's "Full Self-Driving Capability" has many advanced features that begin the path to full automation. They include:

Navigate on Autopilot (Beta): Actively guides the car from a highway's *on-ramp to off-ramp*, including suggesting lane changes, navigating interchanges, automatically engaging the turn signal, and taking the correct exit.

Auto Lane Change: Assists in moving to an adjacent lane on the highway when Autosteer is engaged.

Autopark: Helps with automatically parallel or perpendicular parking the car, with a single touch.

Summon: Moves the car in and out of a tight space using the mobile app or key

Smart Summon: The car will autonomously navigate more complex environments and parking spaces, maneuvering around objects as necessary to find the owner in a parking lot.

Traffic and Stop Sign Control (Beta): Identifies stop signs and traffic lights and automatically slows the car to a stop on approach, with the owner's active supervision.

Upcoming in future models: Autosteer on *city streets,* a much more difficult environment for the system.

For multiple brands, demonstration autonomous cars are now being tested on highways around the United States. The prototypes are easily recognizable. They look like ordinary vehicles but have a laser turret on the roof for determining the vehicles precise position relative to other cars and the edges of the road. And it is done instantaneously and on a continuous basis.

Also visible on the cars are cameras, as well as radar and GPS antennae, jutting out in various spots. These systems provide the sensory input to the onboard "brain", the AI computer.

The "brain" receives the data, makes judgements, and issues commands to the car's controls to guide it on its way. The passengers sit inside in comfortable seats, with no sign of a steering wheel or control pedals on the floor.

Google's Lexus SUV, hybrid, autonomous vehicle

Other manufacturers are working on providing similar systems and are not far behind Tesla.

In 2014, the *Society of Automotive Engineers* published a roadmap to the fully autonomous automobile. They laid out six "Levels" that we will achieve as we move from today's vehicles to fully, self-driving cars.

Level 0 is no automation. Even with blind-spot monitors and forward collision warning, the vast majority of cars on the road today would fall into this category.

Level 1 is adaptive cruise control with radar to automate the distance to the car in front of you, and to automatically brake when that car slows down, and then accelerate as needed. Most luxury vehicles had this type of automation after 2015.

Level 2 includes some automated, lane-centered steering as well as the adaptive cruise control. With this addition, the car will steer itself on a highway using the lane markers as references, but not on normal roads. *Most luxury models in 2021 include Level 2 automation.*

In both Level 1 and 2, the driver needs to still pay close attention while driving because the functionality of these systems is still limited, and the driver may be required to take over at any time.

Level 3: This is conditional automation that constantly monitors the road and can handle most, but not all, driving situations. However, it still requires human intervention if those systems fail. *Many of the automobile manufacturers have said that they will skip this Level and go straight to Level 4.* This is, however, the Level at which Tesla vehicles are now operating.

Level 4: Vehicles at this level are *nearly* self-driving. Level 4 differs from Level 3 primarily because it does not require human intervention if self-driving systems fail. *No driver steering, acceleration or braking controls are required or included in the vehicle.*

Volvo has said it will, however, build a Level 4 XC90 *with redundant driver controls in 2021* because, it says, customers prefer to have the option to drive themselves.

However, other automakers seem reluctant to follow that line of thinking. Ford, for example, has said they will have Level 4 cars on the road in the early 2020's although it's unclear if they will be available to the public or confined to ride-sharing systems, like Uber and Lyft, where driver preferences do not play a role.

Level 5: There are no vehicles currently available at this Level. These vehicles will be *fully self-driving, all the time and everywhere.* They are not driven by humans in any circumstance.

Comparing Level 5 and Level 4, the Level 4 vehicle can drive under limited conditions and works only when those conditions are met. *Level 5, on the other hand, can drive itself under all conditions and just about everywhere.*

Getting to Level 5 is seen by most manufacturers as being difficult. There are some test vehicles underway.

GlobalData, a London-based research firm, forecasts that Level 5 vehicles will begin to appear in extremely low volumes in 2024.[188] Then, there will be a slow ramp-up to around half-a-million units produced in 2030.

That is still only a little more than one half of one percent of the total world market for automobiles. So, we have a long way to go before Level 5 vehicles are commonplace.

Trucking

Trucks are ubiquitous and are a key element in the transportation of goods in the U.S. and the world. One measure of the magnitude of their involvement is that *70% of all commercial freight activity is done by trucks.*

Trucking is a $740 billion a year industry and labor amounts to about one third of that cost.[189] Truck driving is a relatively well-paid profession, although it has its drawbacks. They include spending much of your life on the road rather than at home with your family.

According to the U.S. Census bureau, there are *3.5 million truck drivers* in the United States, making it a major occupation. Many of these jobs, about half, are for the large rigs that require a *commercial driver's license.*

To obtain this certification, the driver must attend an expensive school and take and pass special, government administered exams in order to receive the license.

Yet a significant problem for the industry is that *there is a chronic shortage of qualified drivers.* There has been *an annual, net loss* since of drivers. In fact, more drivers leave the profession each year than those that enter the industry.

There are many reasons, but one may be that the pay is not sufficient to make up for some of the hardships of life as a truck driver. To offset this problem, wages have been increasing, but this is putting pressure on industry profits. *These factors are among the most important forces leading to a push for autonomous trucks.*

As a result, *intelligent trucks may be the first autonomous vehicles* to hit public roads in significant numbers. They can operate around the clock and have lower requirements for human labor.

Most of the large firms engaged in the development of autonomous cars, are also developing truck technology. Trucks need similar approaches as those used in automobiles, but with several additional features.

In fact, a 2019 pilot run by self-driving trucking startup, *TuSimple,* in conjunction with the U.S. Postal Service, showed that autonomous trucks repeatedly arrived ahead of schedule on hub-to-hub routes.[190]

The TuSimple experimental program ran for two weeks. Three of San Diego-based TuSimple's self-driving semis completed five roundtrips. They hauled trailers loaded with mail between USPS distribution facilities in Phoenix, Arizona and Dallas, Texas.

The route was more than 1,000 miles each way. As part of the protocol for the test, each truck had a safety engineer and driver on board to monitor the performance of the autonomous vehicle.

Based on their positive experience with the pilot study, TuSimple now offers an Autonomous Freight Network (AFN) as a service. Currently the network operates between six cities in Arizona and Texas. The company is planning to expand the system, allowing freight to be moved from point-to-point safely and reliably using autonomous trucks.

TuSimple's technology exists because AI hardware and software have reached the level of sophistication that makes the operation possible. Their hardware is based on graphic processor manufacturer, NVIDIA's *"Drive"* platform.[LXII]

This product is *an embedded, **AI**, supercomputing platform that uses Neural Network-based algorithms to process data from camera, radar, and lidar*[LXIII] *sensors.* The system perceives the surrounding environment, localizes the truck to a map, and plans and executes a safe path forward.

Having the capability to process the redundant and diverse information necessary to operate without human supervision, these vehicles are poised to revolutionize delivery and logistics in the near future. *But, so far, only to Level 4 of automation.*

Recall that a Level 4 vehicle can *drive under limited conditions* and work only when those conditions are met. *The manufacturers still have a long way to go to reach Level 5 full autonomy.*

NVIDIA Drive AGX Orin Platform[LXIV]

One of the major car players in the autonomous truck market is *Daimler*, the parent company of Mercedes-Benz. It has been in the autonomous truck race for longer than most of the other

---

[LXII] **NVIDIA** Corporation is a global firm that manufactures graphics processors, mobile technologies, and desktop computers. The company is a leading manufacturer of *high-end graphics processing units (GPUs)*. **NVIDIA** is headquartered in Santa Clara, California.

[LXIII] **Lidar** uses light in a radar system instead of microwaves.
[LXIV] System-on-a-chip with 17 billion transistors, Deep Learning and Computer vision accelerators, capable of 200 trillion operations per second.

competitors. They first demonstrated a self-driving vehicle back in 2014.

Called the *Mercedes-Benz Future Truck 2025*, the vehicle uses a system called *Highway Pilot* to navigate highways without human assistance.

So far, Mercedes has concentrated on *platooning*, where trucks drive themselves closely behind one another, reducing air resistance and lowering their fuel usage by a claimed 10 percent. *Each vehicle still has a driver for safety and for taking over when exiting the freeway.*

Also developing autonomous trucks, along with their cars, are Uber, the ride hailing company, as well as Volvo, Tesla and Google's Waymo.

No developer is now higher than Level 4, and they only have trucks that can operate on major highways with drivers taking over for the non-highway portions of the trip.

Healthcare

Narrow AI applications in healthcare were among the earliest to be developed in any industry. For example, Expert Systems were built in the 1980s to aid physicians in making difficult diagnoses.

Today, AI in healthcare is a large and fast-growing market. Insider Intelligence, the business research firm, projects that this market will grow at an annualized rate of 48%, at least through 2023.

According to *Acumen Research*, the global market for AI in the healthcare industry is expected to rise to *eight billion dollars by the year 2026.*

One area of particular interest to healthcare workers is the *ability to analyze big data sets with AI.* This type of analysis leads to deeper insights into the patients' medical profiles and, therefore, better predictions about health outcomes.

In fact, AI thrives on large amounts of data. The more data, the better the quality of the AI learning and training. While there is an enormous amount of data in most healthcare systems, the problem is that it is presented in a myriad of forms. There may be different data formats, including handwritten notes, which are not easily entered into the system.

The AI algorithms that are being used to analyze the data include *Machine Learning.* It can be used to provide clinical decision support to physicians and hospital staff.

The more sophisticated technique, Deep learning, is now being used to better identify patterns in data and images. This leads to improved analysis of the patient's condition. These applications will be discussed in more detail below.

Medical Imaging

One of the most important areas where AI is impacting healthcare is in *medical imaging.* A key advantage of AI is that it turns images into numbers. Numerical measurements such as volume, dimension, shape, and growth rate can be extracted from the imaging data. That information is placed into databases and then analyzed by the AI to reveal anomalies, patterns, relationships, and insights that human analysis might not see.[191]

There are massive amounts of imaging data available in the healthcare system. IBM reports that radiologists view over eighty-four million studies and ninety billion images each year. Many doctors believe that humans cannot deal effectively with these many pieces of data.

The good news is that *AI thrives on data.* The more images that Machine Learning applications are provided, the more effective their training becomes; and the more potential they have to help clinicians uncover nuances and gain insights.[192]

AI's ability to analyze and quantify imaging data makes it an ideal partner for overworked clinicians. Machine and Deep Learning techniques can discover, isolate, prioritize and report specific anomalies that humans might miss or not initially even consider.

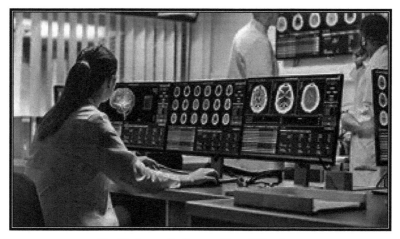

Radiology Information Display

Pathology

Traditional diagnostic methods in pathology are time-consuming and prone to human error. AI technologies, on the other hand, have the ability to handle the gigantic quantity of data created throughout the patient care lifecycle. This ability improves the pathologic classification, prediction, diagnosis, and prognostication of diseases.

Pathologists often spend hours or days handling laborious tasks, such as screening for easily identifiable cancers. These tasks are not necessarily complex, but they are very time-consuming.

The advent of modern, image-based, pathologic, diagnostic techniques is now made possible through the use of AI.

PathAI is a Boston-based company that uses AI to provide pathologists with the cutting-edge tools they need to provide faster, more accurate diagnoses.

Many applications of Machine Learning in pathology use advanced image analysis algorithms to diagnose diseases. Clinicians can process more slides at once and store them as high-resolution, digital files.

Whole-slide imaging and analysis are now possible with AI.[193] The Figure below shows the Machine Learning process which starts at the base of the pyramid and reaches the optimized algorithm at the top.

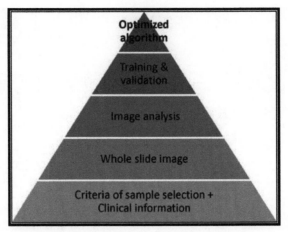

Machine Learning Process for Pathology

AI-based computational pathology[LXV] is an emerging discipline that has shown great promise to increase both the accuracy and availability of high-quality health care to patients. *The most important advantage of computational pathology is to reduce errors in diagnosis and classification while making the pathologist more efficient.*

AI systems have achieved encouraging results with a 92.4% sensitivity[LXVI] in tumor detection rate. In contrast, pathologists, acting on their own, only achieve 73.2% sensitivity. [194]

Clinical Decision Support

Advancements in medical image analysis, such as Radiomics,[LXVII] have expanded our understanding of disease processes and their

---

[LXV] **Computational pathology** is an approach to diagnosis that incorporates multiple sources of data (e.g., pathology, radiology, clinical, molecular and lab operations); uses AI to generate diagnostic inferences; and presents clinically actionable knowledge to users.

[LXVI] **Sensitivity** (True Positive **rate**) measures the proportion of positives that are correctly identified (i.e., the proportion of those who have the condition and who are correctly identified as having the condition).

[LXVII] **Radiomics** (as applied to radiology) is advanced technology that extracts a large number of quantitative features from medical images

management. *But medical professionals can be overwhelmed with all of the data that is now available to them.*

Clinical decision support (CDS) tools are especially helpful when specialists are faced with *a tremendous volume of patient data, and the objective is to reach the optimal, clinical decision.*

CDS tools enhance the specialists' efficiency while flagging potential patient problems and suggesting next steps for treatment.

And now, *AI promises to transform clinical decision-making processes even further. Using the power of AI,* doctors can now harness the vast amounts of genomic, biomarker, and phenotype data that is being generated across the health system. This includes all of the data from health records and delivery systems. The result is an improvement of the safety and quality of patient care decisions.

For example, AI has been incorporated successfully into decision support systems (DSSs) for diagnoses in data-intensive specialties like radiology, pathology, and ophthalmology. Future systems are expected to be *increasingly more autonomous.* They will go *beyond* making recommendations about possible clinical actions to *autonomously* performing certain tasks, such as *triaging*[LXVIII] patients and making screening referrals.

Military

The world has entered an extremely dangerous phase where AI is enabling even small countries to develop Weapons of Mass Destruction like *Slaughterbots.* [LXIX]

---

using data characterization algorithms. The extensive data is then processed with Machine Learning for improved decision support.

[LXVIII] **Triage** – to conduct a preliminary assessment of (patients or casualties) in order to determine the urgency of their need for treatment and the nature of treatment required.

[LXIX] **Slaughterbots** is an arms-control advocacy video produced in 2017. It presents a dramatized, near-future scenario where swarms of inexpensive microdrones use artificial intelligence and facial recognition

Advances in computer power and processing speed are rapidly changing the *scope* of *autonomous* military platforms. In addition, the rapid evolution of advanced algorithms is enabling some amazing capabilities for these systems.

First, some definitions:

- A weapon system that is *"fully autonomous"* is both *self-targeting* and self-mobile.

- A *"drone"* is any unmanned platform operating on land, sea, air, or space. It can be autonomous or controlled by a human.

- *"Drone swarms"* are multiple drones collaborating to achieve shared objectives. These vehicles would be *primarily* autonomously controlled.

The U.S. Air Force is developing *autonomous fighter aircraft* to perform the same missions as now completed by manned aircraft.[195] Not only would pilots be saved from entering dangerous airspace, but the fighter craft is less expensive without the need to accommodate a human onboard.

Before fully autonomous aircraft are ready for deployment, however, developments are underway to *partner humans in current fighter or reconnaissance aircraft to fly with the autonomous drones.* The autonomous partner aircraft will be able to better interpret, organize, analyze, and communicate information without having humans manage each individual task.[196]

Such an autonomous drone has been developed by Boeing. It is a first-generation, sophisticated, autonomous fighter called *"Loyal Wingman"*. It is designed to operate autonomously and fly alongside manned aircraft.[197]

The "Loyal Wingman" drone is thirty-eight feet in length, with twin tails and no space for a human pilot. Essentially, it is a fighter drone that can perform the dynamic movements you would expect from a fighter aircraft. While Boeing is not revealing

---

to assassinate a large number of political opponents based on preprogrammed criteria.

details, it does say that the drone will have a range of about 2,300 miles and be capable of "fighter-like performance."

Given the fast-evolving, *efficacy of modern air-defenses*, a key mission for drones is to scout enemy assets before crewed craft are put at risk. In this application, *fighter-controlled* drones are programmed to fly into heavily defended, or high-risk areas, ahead of the manned-fighter jets. *These drones can also function as a weapons platform to attack enemy targets, like radar installations, prior to the arrival of the crewed aircraft.*[198]

Four "Loyal Wingmen" Accompanying Manned Reconnaissance Aircraft (Boeing illustration)

Drone Swarms

Every part of the US military is also acquiring *drone swarms* for combat missions. Currently it takes a team of people to operate one drone. In the future, one person will be able to control up to one hundred of them.

Distributed Battle Management (DBM) software is being developed to allow networks of staffed and unmanned platforms, weapons, sensors, and electronic warfare systems to interact.

The Navy is also developing *autonomous, unmanned "swarming boats"* that can patrol and defend harbors. This is a high-tech defense against adversaries', like the Iranians, with their manned, low-tech, small attack boats that can threaten Navy ships in the Straits of Hormuz.

The small U.S. boats, called Unmanned Surface Vessels, are designed to conduct Intelligence, Surveillance, Reconnaissance (IRS) missions, find and destroy mines, and launch a range of attacks including electronic warfare, and even firing with mounted guns.

The concept is to use *advanced computer algorithms that bring new levels of autonomy to surface warfare.* This enables ships to coordinate information exchange, operate in tandem without colliding, and launch combined assaults.[199]

U.S. Navy Autonomous Swarm Boat

The Navy technology is called *Control Architecture for Robotic Agent Command and Sensing,* or CARACAS. The approach uses commercial, off-the-shelf AI and sensors to develop systems that are far less expensive than the cost of operating crewed vessels for such routine patrol work.

The AI software enables the autonomous boats to observe the area, detect potential intruder vessels, and decide to follow them.

When a swarm of the autonomous drone boats spots a potential adversary, it dispatches one of their number to approach, while the others continue to track it while continuing their patrol of the area.

"This technology also allows unmanned Navy ships to overwhelm an adversary," said Commander Luis Molina, military deputy for Office of Naval Research's Sea Warfare and Weapons Department. *"Its sensors and software enable swarming capability, giving naval warfighters a decisive edge."*[200]

The *U.S. Air Force has plans to employ swarms of autonomous drones* in a wide range of military roles, from intelligence collection to suppression of enemy air defenses. Greg Zacharias, former chief scientist of the Air Force, believes future F-35[LXX] pilots will incorporate information collected from drone swarms.

Most militaries around the world, including Russia, China, South Korea, the United Kingdom and the United States are *developing swarms of autonomous drones and autonomous weapons systems.* Combining these technologies creates a *slaughterbots-style weapon: an* **armed, fully autonomous drone swarm, or AFADS.** Many people are calling such systems, weapons of mass destruction (WMD) *because of their potential ability to kill large numbers of humans.*

Computer Scientists at the University of Arkansas developed a simulation of such a system.[201]

> *"The system of UAV agents is essentially a hive mind[LXXI] organization controlled by the main loop of the program. However, separate decisions are controlled by the individual UAVs.*
>
> *(In phase 1) Each UAV is instructed in turn to perform a sensor pulse to locate nearby targets. Phase two consists of a negotiation where the UAVs determine their best targets and (communicate with each other to) settle on which UAV should be allowed to track which target.*
>
> *In the third and final phase, the action phase, each UAV is allowed to perform its chosen objective. UAVs are capable of high speed and maneuverability relative to the target vehicles or humans."*

[LXX] The **F-35 Lightning II** is No. 1 in the list of the world's most advanced jet fighters. The aircraft is designed by Lockheed. It is the most flexible, technologically sophisticated, fifth-generation, multirole fighter ever built.

[LXXI] **Hive mind**: is a *unified consciousness or intelligence* formed by a number of individuals, the resulting consciousness typically exerting control over its constituent members.

Let us say that the mission is to terminate a particular individual engaged in terrorist activities. The facial recognition equipped, explosive containing, quadcopter UAV that is selected by the group to target that individual, confirms the identity of the target, and makes a "kamikaze" attack. The resulting explosion terminates the target.

Such a system has been built by the Turkish government. It is called *Kargu* and consists of small, portable, rotary wing, kamikaze drones (see Figure below). The swarm was designed for counter insurgency warfare.[202]

Kargu drones have a facial recognition system, suggesting they can seek out specific individuals. When the autonomous drones swarm, they are too numerous to be challenged by advanced air defense systems. *They can destroy a large number of targets very rapidly.*

The Turkish drone system is just the tip of the iceberg. *The wide availability of the knowledge and equipment to build swarming drone systems* does not bode well for the future. The Center for the Study of the Drone at Bard College (located in Annandale-on-Hudson, New York) has *identified ninety-five countries with military drones.*[LXXII]

Kargu quadcopter autonomous Kamikaze drones

[LXXII] **The Center for the Study of the Drone** at **Bard College** is an interdisciplinary research institution that examines the novel and complex opportunities and challenges presented by unmanned systems technologies in both the military and civilian sphere.

Sophisticated, basic drones can be bought at many retail outlets. Converting them into a deadly swarm only requires the software and hardware to enable the drones to share information and make decisions. And then packing them with a small amount of high explosives completes the deadly package.

With this approach, even small countries with minimal, military resources can add armed, fully autonomous drone swarms, or AFADS, to their arsenals. Like nuclear weapons, if it is possible, then it will happen. *The slaughterbots film warning could become reality indeed.*[203]

Security Applications

Cybersecurity

Cyberattacks are a growing problem, worldwide. Every day, thirty-thousand websites are hacked. The vast majority of global companies have experienced at least one cyberattack. The attacks cost billions of dollars and even poses a national security threat.

Cybersecurity has a wide range of meanings in industry and government. It ranges from individual, home protection systems to the incredible sophistication of the *National Security Agency*'s (NSA) technology. *AI, in all of its forms, is being applied and developed across the entire range of security applications to improve their effectiveness.*

Ransomware is a type of malicious software designed to block access to a computer system until a sum of money is paid. The number of cases in 2020 increased by 150%. They started as attacks on individuals but have now spread to attacks on businesses. *A recent example was an attack, probably from a foreign adversary such as Russia, on a U.S. oil pipeline.*

The Colonial Pipeline incident caused a massive disruption of gasoline distribution in the Eastern part of the United States. The owner of the pipeline finally had to pay millions in ransom to resolve the problem.

The increasing *digitalization* of critical infrastructure sectors such as oil and gas, and the associated industrial systems, is changing the nature of cyber risks. As digitalization drives growth, the energy sector's ecosystem has become increasingly decentralized and complex. *According to the 2021 Global Risks Report, cybersecurity failures are among the top threats facing the world.*[204]

Because of these enormous threats, Cybersecurity, the protection of information technology from malicious attacks, is an area of keen interest at all levels. *AI's capability to analyze massive quantities of data with lightning speed means security threats can be detected in real time, or even predicted based on risk modeling.*

The best approach to stopping the bad guys may be a *combination of focused AI capability, combined with skilled, knowledgeable human partners.* This approach is called Cognitive Security by IBM but is practiced by many firms. The AI part of the Cognitive Security team learns with each interaction to proactively detect and analyze threats, *providing actionable insights for security analysts to make informed decisions.*

Cognitive computing leverages various forms of AI, including Machine Learning algorithms and Deep Learning networks that get stronger and smarter over time. These systems learn with each interaction to connect the dots between threats and provide actionable insights to the analysts.

PatternEx, a spin-off of MIT's *Computer Artificial Intelligence Laboratory,* takes a similar approach. The PatternEx strategy is that Artificial Intelligence, combined with high-performance distributed systems, can augment the skill and intuition of a human Information Security Analyst, at scale and in real time.

PatternEx's *Virtual Analyst* Platform software produces alerts to imminent potential threats. It accomplishes this task by first ingesting data from a large number of sources, computing a broad range of analytics in real-time, and then using a number of *pre-trained AI models* to analyze the data.

Using AI in cybersecurity can be a two-edged sword. Hackers can also use the technology to develop AI-resistant malware. This type of malware can understand the detection patterns used by cybersecurity professionals, *allowing the hackers to penetrate even the most secure solutions.*

Clever Cyber attackers have also learned that they can *target the data used to train the AI cybersecurity models.* The result is that the *accuracy and performance of cyber threat detection systems can be compromised.*

The future of cybersecurity lies with the increasing sophistication of AI-based, anti-malware tools; as well as the next generation of firewalls that learn, detect and defend against new threats as they evolve.

Facial Recognition

Facial Recognition is yet another threat to a free society, made even more ominous by AI.

As the name implies, this *security technology* is aimed at identifying a person based on their photo or video image. The technology first identifies the presence of faces in a photograph and then pinpoints the attributes of those faces. A comparison is made of the facial datapoints in the image to a data base of processed, individual faces in order to make an identification.

Facial Recognition is a relatively old technology. It had its beginnings in the 1960s. Woodrow Wilson Bledsoe was the father of the field when he created the basics of the approach.

Wilson was an American mathematician and computer scientist who is known more broadly for making a number of early contributions to AI in field of pattern recognition.

Woodrow Wilson Bledsoe

With regard to facial recognition, he was the first person to use a RAND tablet, a graphic input device that uses a stylus to electronically mark the location of a point on the tablet. He manually recorded the coordinate areas of facial features like eyes, nose, mouth, and hairline. The manually recorded metrics were saved to a database.

Later, an unknown photograph of an individual, with its standard data points identified, could be entered into the system, and tested against past entries. If that person were in the database, they would be identified. *Wilson's system was the first step taken that proved that facial recognition was a practical biometric.*[205]

This computer-based technology has been available for many years, but only recently has AI been added. The set of Deep Learning algorithms in the AI recognition system looks at such things as eyes open or closed, mood, hair color, as well as the visual geometry of a face. This includes the distance between the eyes, the height of the cheekbones, the distance between the eyes and the mouth, and so on. *AI facial recognition searches on all of those data points and tries to account for variations due to extraneous factors like distance or angle from the camera.*

To be accurate, an AI facial recognition system needs to be trained with a *large amount of facial image data.* The AI algorithms must be exposed to an enormous number of images that vary in ethnicity, age, angles, lighting, and many other factors to develop an effective identification approach.

AI-based software can now instantaneously search databases of faces and compare them to one, or multiple faces, which are detected in a scene. In an instant, you can get highly accurate results. Typically, these systems deliver 99.5% accuracy rates on standard data sets.

However, even well-trained *AI facial recognition systems do not have real-world context and can be fooled.* If you see a colleague who is wearing a face mask, sunglasses, and a baseball cap, *you* may still recognize him. *An AI system, however, might not if it does not have adequate "training".*

Facial Recognition Data Points

AI facial recognition is powerful, *but*, because of its *technical problems*, it *comes with a large set of ethical implications.* This includes the fact that it does *not seem to be possible to regulate the way that facial data for AI systems is harvested.* This means that everyone's right to privacy could be violated at some point.

In spite of these concerns, many countries are forging ahead with the broad usage of AI facial recognition. *China is currently the world leader* in facial recognition applications. They are a totalitarian[LXXIII] society, and facial recognition may be viewed by the government as a major tool in controlling their population.

And there are some ominous indications for all of us about what the future holds for facial recognition technology. We do not perceive it, but there is already a sprawling, invisible, privacy-wrecking surveillance apparatus that surrounds us. It was built under our noses by tech companies, law enforcement, commercial interests, and a secretive array of data brokers and other third parties.

[LXXIII] **Totalitarian:** relating to a system of government that is centralized and dictatorial and requires complete subservience to the state.

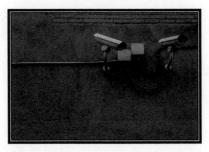

Ubiquitous Security Cameras

It may be hard to believe, but the *FBI's facial recognition database* now includes more than *641 million images* and the identities of an unsuspecting majority of Americans. *This database can be searched by authorities anytime, without warrant or probable cause.*

Throughout most of the United States, facial recognition is legal and almost completely unregulated. It is being used broadly and without our knowledge or permission in such locations as city streets, airports, retail stores, restaurants, hotels, sporting events, churches and, presumably, many other places that we do not know about. And, *as the AI technology versions of the software expands, the amount of data will increase exponentially and be more accessible to additional entities.*

Finance and Banking

AI has had a major impact in the world of Finance. One of the reasons is that an AI system can analyze billions of data points and find patterns and trends. Then the system can use that knowledge to predict the future with much more accuracy than any human. As a result, such capabilities are now finding their places in the areas of Finance and Banking.

For example, with the use of AI technology, many time-consuming banking processes can be automated. AI can "learn" and "memorize" requirements like *Know Your Customer*, anti-money laundering regulations, and the Sarbanes-Oxley Ac. These capabilities ensure that the institutions remain in compliance.

Institutions have massive amounts of data and information relating to the government reporting rules and regulations. The AI system identifies the relevant parts and quickly develops the appropriate reports to meet the law's requirements.

When dealing with large amounts of cash, identity matters. How can you be sure your customers are who they say they are? And you need to know with a high degree of certainty. The use of AI-powered identity verification and authentication solutions have grown dramatically and are now in wide use.

In the Wealth Management sector, AI systems are used to generate individualized, in-depth, wealth status reports for clients. They do it in an easy-to-understand format and much faster than any human could.

Loan management decisions are now highly automated. AI-powered credit scoring applies far more sophisticated rules than traditional reporting systems, helping lenders differentiate between potential borrowers with a high likelihood of defaulting and those that are worthy but have a limited credit history. r

The AI assesses the relevant data of the applicant quickly and efficiently. Payment histories and spending patterns are evaluated, and risks determined. Decisions are reached in seconds rather than months if humans had to do all of the work.

Identity theft and financial fraud are big problems in the Financial Services sector. AI is used to dramatically cut down on these pervasive challenges.

Many of us have received a call from our credit card company. They see charges to our account that they doubt are yours and they want to check in with you.

This type of fraud is identified by the AI system learning the spending habits and behaviors of individual customers. Then it can sense when a transaction in someone's account looks suspicious because it violates the persons normal spending habits. The account can be suspended until the user confirms or denies the transaction.

Digital Assistants

An AI digital assistant is a program that understands voice commands and can perform certain tasks for its user. It operates with *Natural Language Processing* to understand and perform tasks given by clients in natural language. Some companies are working on replacing many tasks now done by humans with such a machine.

Some types of jobs that are going to be replaced include secretaries and administrative assistants that perform tasks such as reading text, taking dictation, finding phone numbers, placing calls, making appointments, checking flight reservations, finding hotels or restaurants, and typing and emailing memos, schedules, and meeting reminders.

An example of this common, relatively low-level Artificial Narrow Intelligence is Apple's Siri, the iPhone technology that answers users' spoken questions. Or Amazon's Alexa that performs the same function.

Alexa, for example, is based on a small, electronic device with a high-quality speaker called Echo. It is located in the home at the convenience of the user. The device is connected to the internet through your home network.

I use Alexa quite often to get useful information like "What time is it?", "What is the local weather going to be today?", "When and where was John Quincy Adams born?". Or to ask it to do simple tasks like, "Play Spotify" or "Turn on the lights".

Actually, you can ask any question you may have. If a browser search can find an answer, then Alexa will get it for you.

"She" can also do things like:

- Find recipes and give you audible step-by-step directions.
- Narrate a Kindle book.
- Get movie showtimes or sports schedules.
- Order pizza or find nearby restaurants.

With Natural Language Understanding (NLU), Alexa can deduce what I actually mean, and not just the words I say. Basically, NLU is what enables Alexa to infer that I am probably asking for a local weather forecast when I request, "Alexa, what's it like outside?

Her responses are provided in the grammatically correct, natural language of an educated, polite, young woman with pleasant sounding, perfectly articulated speech.

She is, however, *not always correct*. Sometimes she does not understand a question and provides the wrong answer. But her accuracy is high enough to make it a very useful application.

The key parameters mastered by a Natural Language AI system like Alexa are the syntax and semantics of the English language. *Syntax is the arrangement of words and phrases to create well-formed sentences. Semantics is the branch of linguistics and logic concerned with meaning.*

Her understanding and use of language are always being improved. For example, if I ask her the time in the early morning, she now immediately replies with the correct answer, then might say, "Have a good morning, Anthony." Shades of Samantha from the movie, "Her"!

"Her" presented the idea of what life would be like if you owned a *Strong* AI version of an Alexa application. Here is the synopsis of the movie from the International Movie Data Base:

> *"In a near future, a lonely writer develops an unlikely relationship with an "operating system" designed to meet his every need. The hero is Theodore, a lonely man in the final stages of his divorce. When he's not working as a letter writer, his down time is spent playing video games and occasionally hanging out with friends. He decides to purchase the new OS1, which is advertised as the world's first artificially intelligent, operating system.*
>
> *'It's not just an operating system, **it's a consciousness,**' the ad states. Theodore quickly finds himself drawn in with Samantha, the voice behind his OS1. As they start spending time together, they grow closer and closer and eventually find themselves "in love".*
>
> *Having fallen in love with his OS, Theodore finds himself dealing with feelings of both great joy and doubt. As an OS, Samantha has powerful intelligence that she uses to help Theodore in ways others hadn't, but how does she help him deal with his inner conflict of being in love with an OS?"*

So, the key differences between the Weak Alexa app and the Strong Samantha app are that *Samantha is conscious and is superintelligent.*[LXXIV] She would easily pass the Turing Test.[LXXV]

An AI system operating at this level would be a huge advance in AI technology compared to today. Currently, *there are no Strong AI applications of any kind in existence.* So, we have years, possibly decades, to wait before we can interact with a Samantha.

Strong AI, or what would be termed Artificial *General* Intelligence (AGI), is defined as the advancement of AI to the point where the machine's intellectual capability is functionally equal to, or surpasses, that of a human.

A Strong AI machine would have the same sensory perceptions as a human being. *To achieve its capabilities, it would use Deep Learning to go through the same education and learning processes as a human child, but on a much faster time scale.*

Humanoid Robots

We have been brought up with images of humanoid robots in many television programs and movies. So, we are familiar with the concept. The reality of today's humanoid robots, however, is a far cry from Commander Data, the android[LXXVI], science officer from the **Star Trek** television programs and movies.

[LXXIV] **Superintelligence** is an intelligence system that rapidly increases its intelligence in a short time, specifically, to surpass the cognitive capability of the average human being.

[LXXV] *The Turing Test is also known as the Imitation Game.* If a machine can engage in a conversation with a human without being detected as a machine, it has demonstrated human intelligence.

[LXXVI] **Android** - a humanoid robot designed to be similar in form to humans. Most humanoid robots are built with the same basic physical structure and kinetic capabilities as humans but are not intended to really resemble people.

Commander Data

Mister Data was treated as a respected member of the crew, but, for most of the series, he was portrayed as being generally devoid of human emotions. He had enough feelings, however, to be unhappy that he was missing this part of being human-like.

The movie, **Bicentennial Man**, is based on a story by science fiction writer, Isaac Asimov. The film, set in the future, follows the 'life' and times of the lead character, Andrew. He is a humanoid robot who is purchased by the Martin family as a servant, programmed to perform menial household tasks. Within a few days, however, the Martin family realizes that they do not have an ordinary robot. There was some kind of "mistake" with his brain. He seems more like a "human" than a robot. Andrew begins to experience emotions and creative thought and becomes a part of his human family. Many years later, he has some alterations and becomes a true android.

Commander Data and Andrew represent the ultimate goal for humanoid robots. But we are still many years away from achieving that level of performance.

Some humanoid robots are currently capable of acting with human-like movements, in fact at the level of a superior athlete. But they do not have the compact, high-performance AI brain of Commander Data or Andrew the Bicentennial Man.

A few years ago, *Atlas*, Boston Dynamics's humanoid robot, could hardly walk. But the company has come a long way in its development of such systems.

Atlas is a research platform designed to push the limits of whole-body mobility. The robot's advanced control system and state-of-the-art hardware give it *the power and balance to demonstrate human-level agility.*

These robots are now capable of remarkable feats, like running, jumping, dancing, and doing back flips. Atlas can even successfully run a Parkour[LXXVII] obstacle course.

Boston Dynamics' Atlas Robot

Atlas, however, only has limited mental capabilities beyond controlling its body. The next step will be to provide it with additional brain power. *Ultimately, the goal for a humanoid robot is to have it "think" and act like a human. This is an active topic of research in AI.*

Theory of Mind

You have heard people say that they "put themselves into the other person's shoes" when they want to understand why that person is doing what they are doing. *Psychologists call this ability "the Theory of Mind".[206]*

It is the power to attribute particular mental states to ourselves and others. It serves as one of the foundational elements for social interaction.

[LXXVII] **Parkour** is a training discipline where practitioners aim to get from one point to another in a complex environment, while running, jumping, and swinging without the assistance of any equipment, and in the fastest and most efficient way possible.

More specifically, it is the ability to know what other people believe or feel or think. And, knowing that information, to *understand that the other person will then act accordingly.*

It is one of the capabilities that differentiates humans from other species. And, up to now, androids or humanoid robots.

The "theory of mind" also explains how people learn to deceive others. First, the deceiver understands that they can say a false or misleading statement to another person in a way that the other person believes the statement to be true. *The deceiver expects that the other person will then act on their false belief and accomplish the deceiver's goal.*

Thus, having the Theory of Mind is the first step to practicing deception. If, for example, young Billy takes a cookie from the jar, eats it before dinner, and gets caught by Mom when she finds the cookie jar open.

"Billy, did you take a cookie!?"

He replies, "Not me. It was the dog!"

Billy is counting on Mom to accept the false belief that Buddy, the dog, was the guilty party. Mom, of course is more experienced than little Billy, so she does not buy into his story.

But, as most parents know, by the time he is a teenager, Billy will become much more sophisticated in his attempts to lead Mom and Dad down the wrong path.

Imagine the implications of a humanoid robot possessing the Theory of Mind, and, therefore, with the ability to deceive humans. That theme was explored in a movie called "Ex Machina".

The chief scientist of a humanoid robot company hires a young computer whiz to "test" the capabilities of the scientist's latest robotic creation. The task is to determine if the female robot could appear to be completely human in her interactions with the young whiz. *To achieve the result, the robot would have to possess the Theory of Mind in order to have the required human-like interactions.*

In the film, the humanoid, female robot passes the test when it successfully uses deception to achieve its own goals, much to the surprise, dismay, and the ultimate destruction of the humans.

The movie is science fiction, but we are at the beginning of a technological development path where such humanoid robot-human interactions will become science fact.

Many scientists around the world are now working toward the creation of highly, human-like robots. Indeed, there are numerous, prestigious universities and institutions engaged in the work of bringing the Theory of Mind to robotics.[207] They include M.I.T.'s Artificial Intelligence Laboratory, the Human Computer Interaction Institute at Carnegie Mellon University, and Bielefeld University in Bielefeld, Germany.

So, what do the humanoid robot designers have to do to create the Theory of Mind in their "progeny"? *To have effective communication between a human and a humanoid robot, there has to be some common ground of understanding.* This includes three things: a shared world knowledge, as well as a shared understanding of social cues, and then the ability to interpret the other's actions.

The overarching capability needed by the robot is to be able to learn from its environment and its interactions with humans. This skill will allow the robot to put all of these capabilities together in the same way that a human would. Research is underway in each of these areas.

Sharing a common world knowledge is the first element. You may have heard of the IBM computer named Watson which has more world knowledge than most humans. For example, it made the Nightly News when Watson beat the best human competitors at playing the TV knowledge game, "Jeopardy".

Actually, Watson is composed of a number of sophisticated computers and associated software, and massive data storage systems all contained in an air-conditioned room.

For a humanoid robot to have this capability, the room full of computers and memory systems would have to be shrunk down to the size of the humanoid robot's brain.

Isaac Asimov in his series, **I, Robot**, called it a "positronic brain". *Scientists believe it will happen eventually, but it is still many years off.*

With respect to the second area, research is underway into understanding human social cues in a way that can be transferred to, and understood by, robots. It is called *Artificial Emotional Intelligence, which will* lead to the humanoid having the Theory of Mind.

Imperial College in London has an ongoing study using AI and a sophisticated vision system to read and understand human emotions.[208] They do this by studying the person's facial expressions. Sensors track the movements of parts of the subject's face and record the emotion displayed at the time.

More than 10,000 faces expressing some emotion, have already been read into the data base, and the system continues to learn how to identify and interpret those human emotions.

The Imperial College researchers' goal is that, someday, this data and information will become part of the brain of a humanoid robot, allowing it to read and understand the thoughts and feelings of its human counterparts. This would be an essential step toward the robot's gaining the Theory of Mind.

To be truly human-like with a Theory of Mind, the humanoid robot would have to be able to put all of these capabilities together and learn from its environment and interactions with humans. *This area of research is called Artificial General Intelligence (AGI).*

A leader in the field is *an English company named **DeepMind**,* based in London, now owned by Google's parent company, Alphabet, Inc. Its mission is to create a set of powerful, general-purpose algorithms that can be used to make a self-learning, AI system. The system is not programmed or made for a special purpose. Instead, like a human, the *Neural Net powered, Deep Learning system* (see the Deep Learning section later in this Part) understands automatically what to do from its experiences.

DeepMind applied this approach to teach its system, AlphaGo, to play the Chinese game of Go.[209] Then a match, consisting of a series of games, was arranged to be played between AlphaGo and the reigning world, Go champion, Lee Sedol.

As simple as the rules of Go are, it is an extraordinarily complex game, much more so than Chess. *There are more possible positions in Go than there are atoms in the universe. That makes Go a googol[LXXVIII] times more complex than chess.*

It would have been impossible to program the IBM Deep Blue computer, which beat Gary Kasparov, to play Go.

AlphaGo's shockingly dominant victory over Sedol in Seoul, Korea, in March 2016 signaled another great leap in the seemingly relentless advancement of machines becoming truly "intelligent" *in the sense of being able to learn enough to outsmart humans.*

So, unlike IBM's DeepBlue and Watson, which were basically tuned to a single application, playing chess, or playing Jeopardy, the machinery behind AlphaGo can be easily adapted to other games and applications.

The general Neural Network / Monte Carlo tree search architecture is a general-purpose tool for games or sequential decision-making under uncertainty.

So, if a humanoid with such capability were to be taught to read human emotions based on the work being done at Imperial College, it could develop a deep understanding of what humans are likely to believe and do in given situations and apply that to any problem it is asked to solve. *It will indeed possess the Theory of Mind. And, therefore, it will have the ability to deceive us.*

More about Strong AI, Artificial General Intelligence, and Artificial Superintelligence later in this Part, and the potential, major benefits to humans. *But they will also present serious potential dangers to humans.*

*The dangers come from the fact that, when we reach Artificial General Intelligence, **there will inevitably be an intelligence explosion which results in the birth of a Superintelligence with potentially unknown and uncontrollable motives that could be harmful to humankind.***

[LXXVIII] A googol is 1×10^{100} or 1 followed by 100 zeros.

The Increasing Power of Computing Technology

Introduction

Since 1956 and the first AI conference, we have made enormous strides in computing power and memory. When I was a student at the Harvard Business School in 1968, we had access to a massive, mainframe IBM computer that was housed in a large, air-conditioned room, cost millions of dollars and was time consuming and difficult to program using punched cards or accessing it through a remote terminal.

Today, *my pocket-size Apple iPhone is 120 million times more powerful in terms of raw computing power than that IBM mainframe, costs a few hundred dollars and has a myriad of easily accessed applications.*

And it is even more astonishing to consider the power of today's IBM mainframes and the complex software available to run on these systems. The massively powerful computers have permitted AI applications to take off. A single IBM z15 mainframe is shown below. It can be integrated into a package of four z 15 systems.

IBM z15 single mainframe

But it is well recognized that *even more immense increases in computing power will be required for Artificial General*

Intelligence (AGI)^LXXIX to be achieved. And such systems are under development.

The modern, electronic computing era can be traced back to the *invention of the transistor* by William Shockley, John Bardeen, and Walter Brattain at Bell Labs in 1947. *A transistor is an electronic, semiconductor component that can amplify or switch electrical signals.* It is used in a circuit to *control the voltage and current efficiently and effectively.*

These devices replaced the big, fragile, power-hungry glass tubes that formerly performed the same functions in numerous electronic devices like radios, television sets and early computers. *The transistors' advantages are that they are more rugged, smaller, cheaper, consume less power and produce less waste heat.*

Transistor Inventors, W. Shockley (seated), J. Bardeen (upper left) and W. Brattain.

In 1959, Bell Labs scientists made another giant leap in the technology, *the metal-oxide-semiconductor, field-effect transistor* (MOSFET or MOS transistor). *It was even faster and more efficient than the earlier transistor designs and was more easily mass produced.*

LXXIX **Artificial General Intelligence** (AGI) is also called strong AI or deep AI. It is a machine and software with a level of general intelligence that mimics human intelligence. The machine has the ability to think, understand, learn, and apply its intelligence to solve any problem, just as humans would do in the same situation.

MOSFET has become the basic building block of modern electronics, and *the most frequently manufactured device in history*, with an estimated *total of thirteen sextillion* (1.3 followed by twenty-two zeros) units produced from 1960 to 2018.

Another big step in computer power, came in 1964. The *General Microelectronics* company introduced the first commercial MOS *integrated circuit (IC)*.[LXXX]

This IC consisted of one hundred and twenty transistors. By 1972, MOS LSI (*large-scale integration*[LXXXI]) circuits were commercialized for numerous applications, including automobiles, trucks, home appliances, business machines, electronic musical instruments, computer peripherals, cash registers, calculators, data transmission and telecommunications equipment.[210]

Modern memory cells were introduced in 1965, when John Schmidt of the *Fairchild Semiconductor Company* designed the first, 64-bit MOS SRAM (Static Random Access Memory).[211]

It needs to be noted that the MOSFET is *also the basis of every microprocessor.*[212] The earliest microprocessors[LXXXII] were built with MOS LSI circuits.

Because of all these developments, the *MOS transistor* has been described as the "workhorse of the electronics industry". It is *the*

basic element of every microprocessor, memory chip and telecommunications circuit in use. [213]

[LXXX] An **integrated circuit** is an electronic circuit formed on a small piece of semiconducting material, usually a silicon wafer, performing the same function as a larger circuit made from discrete components like diodes, resistors, capacitors, etc.

[LXXXI] **Large-scale integration** (LSI) is the process of **integrating** or embedding thousands of transistors on a single silicon semiconductor microchip. LSI technology was conceived in the mid-1970s when computer processor microchips were under development.

[LXXXII] **Microprocessor** - an integrated circuit that contains all the functions of a central processing unit of a computer.

Moore's Law

*In 1965, Gordon Moore, the Director of R&D at Fairchild Semiconductor, and later a cofounder of Intel, made an incredibly significant observation. He postulated that **the number of transistors that can be packed into a given unit of silicon "real estate" doubled about every year**.*

He made this prediction after he had been asked to contribute to the thirty-fifth anniversary issue of **Electronics** magazine with a ten year forecast. His response was a brief commentary entitled *"Cramming more components onto integrated circuits"*. In the article, he said:

> *"The complexity for minimum component costs has increased at a rate of roughly a factor of two per year. Certainly, over the short term this rate can be expected to continue, if not to increase. Over the longer term, the rate of increase is a bit more uncertain, although there is no reason to believe it will not remain nearly constant for at least 10 years."*[214]

Gordon Moore

The idea posited by Moore is that device complexity (higher circuit density at reduced cost) doubles every year. So the relationship would be described as exponential growth.[LXXXIII]

Later, based on actual results, Moore revised down his "law" to *doubling performance every twenty-four months. This is still exponential growth but at a lower rate.*

The actual results over the next forty years are shown on the graph below. It displays the number of transistors on Intel's actual microprocessor chips, starting with the 4004 chip in 1970, to the Xeon 5500 chip in 2010. The number of transistors per processor chip did indeed double every twenty-four months over this forty-year period.

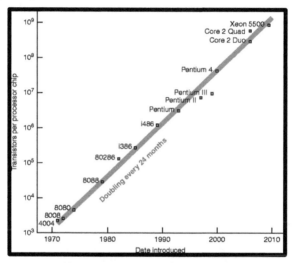

Moore's Law, Transistors used in Intel chips (1000's) vs. Time

But there is a limit to how long Moore's Law will hold. It has to do with the physical limitations of the manufacturing process. *As the scale of chip components gets closer and closer to that of*

[LXXXIII] **Exponential growth** is a pattern of data that shows greater increases with passing time, creating the curve of an exponential function. When plotted on a log-log scale, as shown in the Figure, the function is a straight line.

individual atoms, it has become more difficult and expensive to keep up the pace of Moore's Law.

Industry experts have not reached a consensus on exactly when Moore's law will cease to apply, but everyone agrees it is close. Microprocessor architects report that semiconductor advancement has slowed industry-wide since about 2010. It is now below the pace predicted by Moore's law.[215]

The industry is, however, embracing other kinds of technologies to keep increasing computing performance. This includes advanced software frameworks and tools, as well as new ways of packaging the chip circuitry. In the near future, a whole new computing technology, quantum computing, will change the paradigm once again (see the section ahead, *Quantum Computing*). [LXXXIV]

The Law of Accelerating Returns

Moore's Law, is just one manifestation of *the greater trend that all technological change occurs at an exponential rate.* So says Ray Kurzweil, an American polymath, who is an expert in computer science, a prolific inventor, author, MIT professor, founder of companies, corporate executive, and a futurist. He is now the Director of Engineering for Google where he is responsible for the company's AI research program.

Ray Kurzweil

Kurzweil calls this exponential growth of technology *the law of accelerating returns.* He believes it applies to many human-

---

[LXXXIV] *A paradigm shift* is defined as an important change that happens when the usual way of thinking about or doing something is replaced by a new and different way.

created technologies such as computer memory, transistors, microprocessors, DNA sequencing, magnetic storage, the number of Internet hosts, Internet traffic, decrease in device size, and nanotech citations and patents. He has demonstrated and documented his thesis in his many speeches and writings.[216]

Whenever a technology approaches some kind of a barrier, according to Kurzweil, a new technology will be invented to allow us to cross that barrier.

According to Kurzweil, *Moore's Law is actually the fifth in a series of paradigms*[LXXXV] *of exponential computing power growth.* The previous four paradigms include—"the mechanical calculating devices used in the 1890 U.S. Census, Turing's relay-based machine that cracked the Nazi enigma code, the CBS vacuum tube computer that predicted the election of Eisenhower, and the transistor-based machines used in the first space launches".[217]

All of these machines operated under the same underlying principle of exponential growth, says Kurzweil. *In other words, even if scientists do hit a wall with silicon-based circuits, something will emerge to take its place.* **So, he believes that the growth of computing power will continue to increase at an exponential rate.**

Supercomputer Technology

Many scientists believe that machine superintelligence and consciousness will emerge as soon as the power of computers reaches a critical level, much higher than current capabilities. Others do not share this optimism.

Today's most powerful computers used in AI research are called supercomputers. Essentially, they consist of an exceptionally large number of off-the-shelf computers running in parallel. As the name implies, they are able to operate at a much higher level, performing enormous numbers of calculations per second compared to general purpose computers.

[LXXXV] A **paradigm** is a standard, or a perspective, or set of ideas. Essentially, a *paradigm* is a way of looking at something. ... When you change paradigms, you're *changing how you think about something.*

These powerful systems had their beginning in the 1960s when Seymour Cray (1925-1996) of Control Data Corporation developed the first model. Cray was an American electrical engineer and supercomputer architect who designed a series of computers that were the fastest in the world for decades.

Later, he founded *Cray Research* which built many of these machines for special applications. Because of Cray's pioneering efforts, he is called "the father of supercomputing", and has been credited with creating the supercomputer industry.

Joel S. Birnbaum, then chief technology officer of Hewlett-Packard[LXXXVI], said of Cray:

> *"It seems impossible to exaggerate the effect he had on the industry; many of the things that high performance computers now do routinely were at the farthest edge of credibility when Seymour envisioned them."*

Seymour Cray

[LXXXVI] Hewlett-Packard acquired Cray, Inc. in September 2019.

Massively parallel[LXXXVII] supercomputers have become the norm to achieve this performance.[218]

A Cray Supercomputer

The performance of a supercomputer is commonly measured in floating-point operations per second (FLOPS). Today's fastest supercomputers operate in the range of 100 Peta FLOPS, or 100×10^{15} FLOPS.

Top Twelve Fastest Supercomputers in 2020

Number	Name	Manufacturer	Speed (PetaFLOPS)
12	Sequoia	IBM	17.12
11	Pangea III	IBM	17.8
10	Lassen	IBM	18.2
9	SuperMUC-NG	Lenovo	19.4
8	AI Bridging Cloud Infrastructure	Fujitsu	19.8
7	Trinity	Cray	21.2
6	Piz Daint	Cray	21.2
5	Frontera	Dell EMC	23.5
4	Tianhe-2A	NUDT	61.4
3	Sunway TaihuLight	NRCPCE&T	93
2	Sierra	IBM	94.6
1	Summit	IBM	148.6

[LXXXVII] **Massively parallel** is the term for using a *large number* of computer processors (or separate computers) *to simultaneously perform a set of coordinated computations in parallel.*

In 2020, the fastest supercomputer in the world was *Summit,* an IBM design located at Oak Ridge National Laboratory in the United States. IBM was the major manufacturer of supercomputers in 2020, with five machines in the top twelve in the world. In 2021, some newer, faster machines came online.

Cray, Inc., still a major player[LXXXVIII]*, with numbers six and seven in the top twelve,* has a $600 million contract to build "Frontier". It will be installed in late 2021 at Oak Ridge National Laboratory and be able to perform at a speed of 1.5 exaFLOPS, or 1.5×10^{18} FLOPS. That makes Frontier about ten times faster than Summit, also at Oak Ridge, which was the fastest supercomputer in 2020.

Numbers 3 and 4 on the supercomputer list are Chinese in origin and are located in that country. They were manufactured by Chinese government agencies and are used in their defense establishment.

The operating system for modern supercomputers tends to be based on Linux.[LXXXIX] Each manufacturer has its own specific Linux-derivative. This is because the differences in hardware architectures require changes to optimize the operating system to each hardware design.[219]

Quantum Computing

As Ray Kurzweil has said in his *Law of Accelerating Returns,* when one technology area peaks in its ability to produce exponential increases in performance, another one springs up to

---

[LXXXVIII] **Cray Inc.** was formed in 2000 when Tera Computer Company purchased Cray Research Inc. and adopted the name of its acquisition. The company was then acquired by Hewlett Packard Enterprise in 2019 for $1.3 billion.

[LXXXIX] **Linux®** is an open-source operating system (OS). An operating system is the software that directly manages a system's hardware and resources, like CPU, memory, and storage. The OS sits between applications and hardware and makes the connections between the applications and the physical resources that do the work.

take its place. As Moore's Law starts to reach that point in standard computer technology, *it may be Quantum Computing that will take over to continue the exponential growth of computing power.*

Today's state-of-the-art AI consists of many narrow, functional applications that are limited by the computational capabilities of standard computer systems.

Quantum computing and artificial intelligence are both paradigm shifting technologies, and artificial intelligence is likely to require quantum computing to achieve significant progress toward achieving *Artificial General Intelligence (AGI).*

The implications of Quantum Physics for Computing

Clearly, Quantum Physics is vastly different from classical physics. The result is that the more we explore the implications of quantum physics in our reality, the weirder and more mysterious it seems.

You may not recognize it, but quantum physics has had a major impact on our lifestyles. Many of the advanced technology products we use today, like computer chips and lasers, have resulted from our understanding of quantum physics. And now we are on the verge of dramatically increasing our computer power with a revolutionary approach that depends even more directly on quantum properties.

Qubits, instead of bits, are the fundamental building blocks of a *quantum computer.* They also come into play for one theory of how consciousness occurs proposed by acclaimed physicist, Roger Penrose. But more about that later in this section.

One of the quantum properties used in qubits is called "superposition". Instead of just a one or a zero, as the case with the bits on a computer chip, a qubit can be in multiple states *at the same time,* as described by the Schrödinger wave equation.

It's hard to believe, but a qubit can be both a one and zero simultaneously! An analogy to the idea of quantum superposition is when you pluck two strings of a guitar at the same time. The sound waves you hear are the combination of the two sounds, or what is termed, a "superposition" of the two notes.

How is the phenomenon of superposition even possible in the qubit? It is precisely because quantum systems, such as electrons, behave as waves, like the soundwaves in superposition of the two notes coming from the guitar. In fact, quantum physics says that an *electron's normal state is a wave in superposition of all its possible properties.*

The term "quantum coherence" contemplates the situation described in the double slit experiment, discussed above, where a laser light wave is split in two by the slits. The two waves, emerging from the adjacent double slits, *coherently* interfere with each other.

Because of their common origin, a coherent laser beam, the two waves are coherent, meaning that their frequency and waveform are identical, and their phase difference is constant.

More generally, the idea of *coherence* applies to the *maintenance of all the physical properties in multiple quantum systems.*

Quantum decoherence, on the other hand, is the *loss* of quantum coherence. Qubits, the basic element of a quantum computer, require that the quantum system maintains its coherence to operate properly. If decoherence occurs, the necessary properties of the qubit, such as superposition, soon disappear. *And the computer would become inoperative.*

If a quantum system were perfectly isolated, it would maintain coherence indefinitely. When it is not perfectly isolated, for example during a measurement, the qubit is affected by the environment and its coherence is lost over time.

It is important to note that decoherence represents a challenge for the practical realization of quantum computers. Current qubits are made with superconducting materials. To maintain their quantum states, they must be kept at close to absolute zero, which is minus 460 degrees Fahrenheit. Otherwise, thermal energy causes interference and decoherence.

Just as strange as a wave being the superposition of all of its properties, is the quantum phenomena of *entanglement. This is another form of superposition.*

Interestingly, physicists agree that quantum coherence and quantum entanglement are operationally equivalent, as two kinds of superpositioning.

Here is how it works. Two quantum systems, like those found in a qubit, can interact, and become entangled. That is, they are inextricably linked in perfect unison. *A change in one quantum system results in an instantaneous, related change in the other.*

But even more mystifying is that they *continue to be linked* in this *way no matter how far they are separated*, **even as far apart as opposite ends of the universe!**

The observation of entanglement really bothered one of the most famous physicists of all time, Albert Einstein, creator of the Theory of Relativity. Einstein is also one of the founders of quantum physics. In fact, his only Nobel prize was for quantum physics, not relativity. *What bothered Einstein was a conflict between the findings of the two fields of relativity and quantum physics.*

One of the major results of his relativity work was to show that the speed of light is the speed limit for anything in the universe. This is approximately 186 thousand miles per second. At the speed of light, it would take 93 billion years for the two entangled particles at opposite ends of the universe to communicate with each other! Yet, they do so, instantaneously! As a result, Einstein termed entanglement to be "spukhaften fernwirkung" or "spooky action at a distance".

Max Planck was a German theoretical physicist that made a fundamental contribution to quantum mechanics. Some regard him as the father of Modern Physics. It was because of his work that physicists started to think about the idea of the quantum.

He developed his concept *to solve the problem in physics* at that time called *The Ultraviolet Catastrophe.*

The *ultraviolet catastrophe* was the *prediction* of physicists based on their current model of how a black body[XC] behaved. *They*

[XC] **A black body** is an idealized, physical body that *absorbs* all incident electromagnetic radiation. Since no light is reflected, it appears to be perfectly black. A black body also *emits black-body radiation.*

believed from their mathematics that a black body would instantaneously radiate all of its energy until it was near absolute zero. This is what they termed to be a "catastrophe".

But the physicists' *experiments* showed that this *catastrophe did not occur.* Since their current understanding was obviously incorrect, *a new model for the behavior of black bodies was needed.*

So, Planck was studying how black body's radiate energy. Based on his observations, he formulated the Planck postulate. *Electromagnetic energy is emitted only in quantized form, or in chunks rather than continuously.*

The energy in a "quantum" of radiation was given by Planck's constant, h, times the frequency of the radiation, v. Physicists now call these quanta of light, *photons.* He won the Nobel Prize in Physics in 1918 for this work.

Later, Planck surprised a lot of people when he said in a 1931 interview with the Observer (a British Sunday newspaper):

"Consciousness, I regard, as a fundamental. I regard matter as derivative from consciousness. We cannot get behind consciousness. Everything that we talk about, everything that we regard as existing, postulates consciousness."

Planck may have been reflecting on the results of the double-slit experiment that showed that factors associated with consciousness and our physical, material world are, in some way, connected. *Some physicists are being forced to admit that consciousness may play a fundamental role in the universe.*

The First Quantum Computers

The first discussion about using quantum phenomena for computing applications was in 1980.[220] Paul Benioff, a physicist with Argonne National Laboratory, proposed a quantum mechanical model of the Turing machine. Serious research into quantum computers did not take off, however, until about thirty years later. And it is still in its infancy.

Quantum computing makes use of the very strange and mysterious quantum phenomena of superposition, and entanglement to perform its data operations.

Qubits, instead of bits, are the fundamental building blocks of a quantum computer. One of the quantum properties used in qubits is called "superposition". Instead of just a one or a zero, as is the case with the bits on a standard computer chip, *a qubit can be in multiple states at the same time.* While it is difficult to comprehend, *a qubit can be both a one and zero simultaneously. This essentially lets each qubit perform two calculations at once.*

Dr. Paul Benioff

An analogy to the idea of quantum superposition is when you pluck two strings of a guitar at the same time. The sound waves you hear are the combination of the two sounds, or what is termed a *superposition of the two notes.*

How is the phenomenon of superposition even possible in the qubit? It is precisely *because quantum systems, like electrons, behave as waves,* similar to the soundwave in superposition of the two notes coming from the guitar. In fact, *an electron's normal state is a wave in superposition of all its possible properties.*

Electrons behave as waves, that is, until physicists observe them during an experiment. Then the electrons suddenly change and behave like particles. Yes, that is correct. *An electron is a wave, except when it is observed by a human, and then it turns into a particle!*

This sounds incredible, but the electron seems to know it is being watched and the wave function, which describes its normal behavior, "collapses", and the electron then acts like a particle.

Now, that is bizarre. *This understanding of quantum physics is called the Copenhagen Interpretation and is agreed upon by most physicists **because it works to describe reality well**, but, frankly, is still not completely understood.*

Just as strange is the quantum phenomenon of *entanglement.* Two quantum systems, like those found in a qubit, can interact, and become entangled. That is, *they are inextricably linked in perfect unison.*

A change in one quantum system results in an *instantaneous,* related change in the other. But even more mystifying is that they continue to be linked in this way *no matter how far they are separated physically, even as far apart, theoretically, as opposite ends of the universe!*

This observation really bothered one of the most famous physicists of all time, Albert Einstein, creator of the Theory of Relativity. Einstein is *also* one of the founders of quantum physics. In fact, his only Nobel prize was for quantum physics, not relativity.

What bothered Einstein was a conflict between the findings of the two fields of relativity and quantum physics.

One of the major results of his relativity work was to show that *the speed of light is the speed limit for anything in the universe.* This is approximately 186 thousand miles per second.

At the speed of light, it would take 93 billion years for the two entangled particles at opposite ends of the universe to communicate with each other! Yet, they do, instantaneously! As a result, Einstein termed entanglement to be "spukhaften fernwirkung" or *"spooky action at a distance".*

IBM and Google are now two leading companies working to develop quantum computing technology. In October 2019, Google AI, in partnership with NASA, claimed to have demonstrated *"quantum supremacy"*[XCI] by performing a quantum computation

---

[XCI] **Quantum supremacy**: In quantum computing, *quantum supremacy* or quantum advantage, is the goal of demonstrating that a programmable quantum device can solve a problem that no classical computer can solve in any feasible amount of time.

that was infeasible on any classical computer *including supercomputers.*[221,222]

Google's Quantum Computer removed from its tank

The promise of quantum computing is that certain computational tasks can be executed exponentially faster on a quantum processor than on a classical processor.

The Abstract[223] to a 2019 article in **Nature** on achieving quantum supremacy reads:

> *"Our Sycamore (quantum) processor takes about 200 seconds to sample one instance of a quantum circuit a million times—our benchmarks currently indicate that the* **equivalent task for a state-of-the-art classical supercomputer would take approximately 10,000 years.** *This dramatic increase in speed compared to all known classical algorithms is an experimental realization of quantum supremacy for this specific computational task, heralding a much-anticipated computing paradigm."*

While a remarkable accomplishment, achieving *large scale quantum computing with the technology used in this experiment is highly problematic.* Most quantum computers today rely on qubits based either on superconducting circuits or trapped ions. A major drawback with these approaches is that they *demand operating temperatures colder than those found in deep space, about negative 456° F.*

The reason is that thermal vibrations can disrupt the qubits. At higher temperatures, the quantum states of the qubits become decoherent and lose their quantum properties.

The expensive, bulky refrigeration systems required to hold qubits at such frigid temperatures make it an extraordinary challenge *to scale these platforms up to high numbers of qubits.*

But such scaling will be necessary to reach interesting and useful levels of computing power.

So what is *needed is a quantum computer that operates at room temperature.* A little over a year after the experiment reported in **Nature**, a team based in China took this achievement a step further. They carried out their work on *a photonic, quantum computer **working at room temperature.***

In this system, infrared, laser pulses are fired into a chip and are coupled together with microscopic resonators to generate so-called *"squeezed states"* **consisting of superpositions of multiple photons.**

The photons next flow to a series of beam splitters and phase shifters that *perform the desired computation.*

With this approach, the Chinese team found that their photonic, quantum computer generated solutions to the boson-sampling problem[XCII] in 200 seconds. They estimate these *solutions would take 2.5 billion years to calculate on China's TaihuLight supercomputer. This implies a quantum advantage of 10^{14}.* [224]

A programmable, photonic circuit has been developed that can execute various quantum algorithms. It is potentially highly scalable[XCIII]. *Such a device could pave the way for large-scale, quantum computers based on photonic hardware that operate at room temperature.*

[XCII] **Boson sampling problem** - **bosons** are particles that act as force carriers. They function as the 'glue' holding matter together. "**Boson sampling**" is a computational problem. It consists of sampling from the output distribution of indistinguishable **bosons**.

[XCIII] **Scalability** is the ability to increase the size and power of the quantum computer using the same photonic technology.

For example, when quantum computers reach 50-100 qubits, they will be able to perform tasks that surpass the capabilities of today's classical digital supercomputers.

If two qubits are quantum-mechanically linked, or entangled, they can perform 2^2 or four calculations simultaneously; three qubits, 2^3 or eight calculations; and so on. So, 75 qubits could perform 2^{75}, or $3.8x10^{22}$ calculations simultaneously. *This is far faster than the most advanced of today's supercomputers.*

Looking further into the future, a photonic quantum computer with only 300 qubits could perform more calculations in an instant than there are atoms in the visible universe.

The photonic systems can also readily integrate into existing fiber optic–based, telecommunications infrastructure, potentially helping connect quantum computers together into powerful networks and even a quantum Internet.

*With the addition of so-called **time multiplexing**[XCIV] architecture, photonic quantum computing could, in principle, scale up to millions of qubits.[225]* **But there are still considerable technical difficulties that have to be overcome to reach those levels.**

There are numerous, small companies working to overcome these problems and develop photonic integrated circuits for quantum computer applications. One such company is *Xanadu Quantum Technologies*, a Canadian firm based in Toronto. It has some partnerships with larger firms like IBM's Q network and Amazon's Quantum Solutions Lab. The small company is offering *the first photonic quantum computer on the cloud.[226]* It has also raised $100 million in new funding as investor interest in the quantum photonics industry heats up.

With this promise, photonic quantum computers could well provide the means for AI researchers to develop and implement widespread, Artificial General Intelligence in the foreseeable future.

[XCIV] **Time multiplexing** is a sharing technique. Specifically, it breaks up the time you have available into a stream of fixed-sized slots and distributes those slots among the various activities that need to be accomplished.

Cloud Computing

Cloud computing is the delivery of various *internet-based*, IT services. Such tools include data storage, servers, databases, applications, and networks. *Much of the growth of AI applications will come from Cloud-based systems.*

Cloud-based digital assistants such as Apple's Siri, Amazon's Alexa, and Google Home combine a seamless flow of artificial intelligence technology and cloud-based computing resources to enable users to have questions answered, make purchases, adjust a smart thermostat, or hear a favorite song instantly.

Yet these applications are just the tip of a massive iceberg that is cloud computing. Since it takes a great deal of processing power to run advanced artificial intelligence algorithms effectively it has been extremely cost-prohibitive for most companies to deploy AI. But now *the integration of AI on cloud computing platforms changes the game. The cost of AI computing power can now be shared among many users, making it cheaper for everyone.*

Software-as-a-service providers are now including new *cloud-based AI tools* in larger software suites to provide greater functionality to end-users. For example, the popular Customer Relationship Management (CRM) platform, *Salesforce*, recently *added Einstein. Salesforce Einstein is* an AI tool that offers the ability to *capture and analyze customer data.* This makes it easier to track and personalize customer relationships. Salesforce Einstein is the first comprehensive AI for a CRM.

Microsoft Azure products and services are examples of the capabilities currently being offered for *cloud-based AI applications* by mainstream firms like Microsoft. Their AI technologies allow customers to analyze images and videos, comprehend speech with NLP, and make predictions using Deep Learning data analysis. Knowledge-mining AI develops hidden insights from customer's business data. These kinds of services allow even small companies without the necessary programming and hardware capabilities to take advantage of today's AI platforms.

AI and The Internet-of-Things

If you are a contemporary homeowner, you are probably familiar with the Internet-of-Things (IoT). Internet connected, "Smart" items in your home may include i-phones, fitness watches, appliances, flat screen televisions, i-pads, thermostats, and home security systems with cameras.

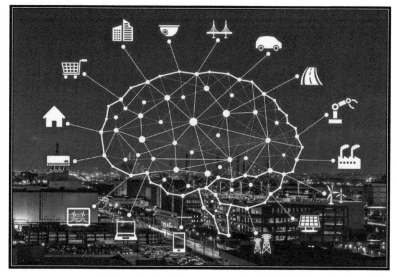

AI and the Internet-of-Things

All of these IoT devices use the internet to communicate, collect, and exchange information about our online activities. As a result, every day, *these devices generate about one billion Giga Bytes of data.*

By 2025, there is projected to be a large increase to *forty-two billion IoT-connected devices globally.* The implication is, therefore, that *there will be an enormous growth in the amount of data generated by these products and* **AI will be necessary to collect and use it effectively.**

AI powered, Deep Learning[XCV], for example, will help to use the mass of IoT connected records. The AI will simulate smart behavior, with little or no human intervention, to "mine" the data and then analyze it for users.

With this AI integrated approach, *Smart homes* are able to control appliances, lighting, and other electronic devices, while *learning* a homeowner's habits and developing automated "support" for the household. One obvious side benefit of this level of automation is improved energy efficiency.

The smart thermostat solution manufactured by *Nest* is a good example of home, AI-powered IoT. The smartphone integration with the thermostat allows the system to check and manage the temperature from anywhere, inside or outside the home.

NLP (Natural Language Processing) is getting better at allowing people to communicate more easily with their IoT devices. *Alexa and Siri are good examples of this capability.*

It is becoming clear that pairing AI with IoT will *also help businesses to better understand a broad range of risks in their enterprises, and then provide a prompt response to problem situations.* These risks include financial loss, employee safety hazards, and cyber threats.

For example, Fujitsu helps ensure worker safety by engaging AI to analyze *data sourced from the employee connected, wearable devices, looking for potential threats. Necessary actions can then be taken to mitigate the problems.*

Because of these skills, both startups and large companies see AI technology as unleashing the full potential of their IoT. So, the leading vendors of IoT platforms like Oracle, Microsoft, Amazon, and Salesforce have started consolidating AI capabilities into their IoT applications.

[XCV] **Deep learning**: a subset of Machine Learning, which is essentially a Neural Network with three or more layers. These Neural Networks attempt to simulate the behavior of the human brain, although at a very low level, allowing it to "learn" from large amounts of data.

Also outside of the home, there are several sectors of the economy where connected, "smart" products are having a major impact. Two of these sectors are:

- Healthcare
- Transportation

In the healthcare sector, remote health monitoring is a significant trend. Examples are "at-home" blood pressure and heart rate monitors for patients with those risk factors. The home monitors are able to connect to the hospital over the internet and periodically report their data to the doctors.

If a heart patient has a pacemaker implanted, it is critical for the clinic to monitor its performance on a regular basis. This can now be done remotely with a hand-held device that connects the pacemaker and an i-phone using Bluetooth technology, and then to the internet using the i-phone. The detailed performance information from the pacemaker is sent automatically to the device clinic in the hospital for their analysis and evaluation. Any necessary adjustments can then be made.

In the *field of transportation*, the Internet of Things is used to link the vehicle to the transportation infrastructure, and then to the driver to make the combination operate more efficiently and effectively. *Self-driving cars* are an application that is currently in heavy development and AI and IoT will make significant contributions to this advance.

Vehicle-to-everything communications is one of the first steps to achieving the goal of autonomous driving and connected road infrastructure. Autonomous driving will be covered in detail later in this Part.

The AI in Tesla's self-driving cars predict the behavior of pedestrians and other vehicles under various circumstances. The AI can also determine road conditions, optimal speed for the vehicle, and the impact of weather conditions. The ability to get smarter with each trip is built into the Tesla AI.

Retail stores use customer, data gathering and analytics to reduce checkout time and increase the productivity of the cashiers. The system has numerous cameras and sensors throughout the store to obtain data points on customers' movements. Then the analytics predicts when those customers will reach the checkout line. The program then suggests dynamic staffing levels for the checkouts in order to obtain the efficiency point sought by the retail store management.

Computer Vision

There are many related peripheral technologies that will be required for Artificial **General** Intelligence (AGI) or human-like AI. One of the most important of these is a vision system to provide sight for the AGI. The technology is already well-developed and in wide operation.

Computer vision, or image recognition software, is a subset of Artificial Intelligence. It allows computers to see and make sense of the world. With this technology, a computer can learn to analyze photos, videos, or thermal and infrared data and then, based on that information, develop a clear understanding of its environment or situation. With this knowledge the computer can make appropriate decisions about what to do.

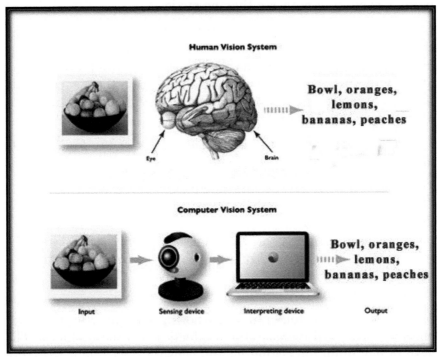

Pictured above is a comparison of a Human Vision System and the analogous Computer Vision System (CVS). The object to be identified is a bowl of fruit seen on the left. The human's eye is the visual sensing device that captures the image of the fruit bowl. The brain interprets the visual information from the eye as a bowl of fruit and that is conveyed as the output to the conscious person.

With the Computer Vision System, the visual sensing device is a sophisticated camera. The computer-based AI software interprets the visual data, identifies the bowl of fruit, and transmits the information to the system output.

The accuracy of object identification in a Computer Vision System has increased to an incredible 99%. This makes computer vision more accurate than humans at seeing and reacting rapidly to visual data. This is a very necessary capability for autonomous driving of vehicles and other vision critical applications.

It is important to understand the difference between Computer Vision and *Machine Vision systems. As discussed, computer vision technology* is used to *automate image processing. A Machine vision system, on the other hand,* uses *computer vision in real-world interfaces, such as a factory line to perform a function like automated inspection(see* **Intelligent Maintenance/Automated Product and Part Inspection,** *later in this Part).*

Machine Learning

Our laptop and desktop computers use application programs, like word processors and spreadsheets, to manipulate verbal and numerical data in order to do what we want to get done. These programs are powered by the instructions written into them by humans.

Many AI applications, however, use a different approach called Machine Learning (ML). **This form of AI is able to learn and adapt without following explicit instructions**. Instead, it uses algorithms and statistical models to analyze and draw inferences from patterns in data and *teaches itself* how to make the desired decisions. For example, ML systems are used to :

- Categorize or catalog items like people or things,
- Predict likely outcomes or actions based on the identified patterns,

- Detect anomalous or unexpected behaviors.

This approach is more like the process that humans use in their learning activities. Accuracy is gradually improved with time and experience as the training progresses.

There are multiple approaches to Machine Learning that vary in their complexity and the level of sophistication involved.

The "classical" approaches to Machine Learning include:

- *Supervised learning*
- *Unsupervised learning*
- *Reinforcement learning*

We will consider each of these areas and compare their applications. **Deep Learning** is the latest and most sophisticated Machine Learning technology. It will be covered in detail later in this section.

Machine Learning had its beginning in 1959 at an IBM research laboratory. One of their researchers, Arthur Samuel, developed a Machine Learning-training program for an IBM 7094 computer to teach it to play checkers.[227]

When the training was complete in 1962, Samuel arranged for the computer to play Robert Nealy, a master checkers player, to establish the system's level of expertise.

The computer won the match. *This was the first time that a computer beat a human at such a game, and this achievement became a well-known milestone in AI history.*

Arthur Samuel

Early Machine Learning approaches depended on some degree of human intervention. With ***Supervised Learning,*** experts determine the set of features to be studied in designing the algorithm[XCVI].

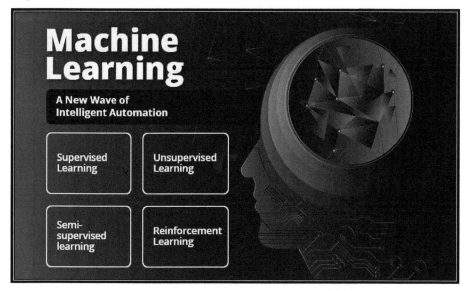

For example, if you want to forecast Gross Domestic Product (GDP) as the levels of employment and income change, you need to use a data set that contains many years of actual results for the dependent (GDP) and independent (employment and income levels) variables.

The computer needs to distinguish the differences between the data points, so a *structured data set* must be provided. The algorithm uses this data to determine correlations and logic that can then be used to predict the desired answer, in this case, GDP.

As more and more data regarding recorded results are provided to the "machine," the algorithm's performance improves in terms of accuracy of the algorithm's output. *As a result, the machine's "intelligence" increases over time.*

[XCVI] An **algorithm** is a set of instructions designed to perform a specific task. This can be a **simple** process, such as multiplying two numbers, or a complex operation, such as playing a compressed video file.

I first started to use this technique for micro economic forecasting back in the 1970's, using a terminal to access an IBM mainframe computer. I built multiple regression models with a Machine Learning algorithm. The regression results produced a predicted value based on the independent variables. *Nowadays, you can just go to your Excel spreadsheet program and find a regression function that will easily do the calculations for you.*

There are four parts to developing a classical Machine Learning algorithm, of the type described above.

1. *Design:* The first step is to identify the problem and thoroughly understand it.

2. *Decision Process*: Using some initial data points (part of the larger training data set), an algorithm is generated which will produce an estimate of the desired outputs based on an identified pattern in the data.

3. *Error Function*: To *evaluate the accuracy of the predictions of the model,* additional examples from the larger data set are used. The error function *compares the model's predicted values to the actual values of the known examples.*

4. *A Model Optimization Process*: to make the model better fit the data points in the total training set, *weights are adjusted to reduce the discrepancies between the known examples and the model's estimates.* The algorithm will repeat this "evaluate and optimize" process, updating weights autonomously until a desired threshold of accuracy has been met.

It is important to understand the limitations of this approach. First, the output of a classical Machine Learning algorithm is entirely dependent on the data to which it is exposed. Change the data, and you change the result.

Furthermore, the algorithm's output is *not autonomously creative.* The machine will not spontaneously develop new hypotheses from data not in evidence. Nor can the machine determine a new way to respond to emerging changes.

Unsupervised learning *is another approach to Machine Learning.* It is used when the there is *a large amount of unlabeled data,* as found in social media applications.

Understanding the meaning behind this data requires algorithms that classify the data based on the patterns or clusters it finds. The computer uses an iterative process, analyzing the data without human intervention.

A common application is for a spam-filter where Machine-Learning classifiers, based on clustering and association, are applied to identify the unwanted emails.

A more sophisticated version of "supervised learning" is ***"reinforcement learning"****.* This is the general approach taken by Arthur Samuel in teaching the IBM computer to play checkers.

Reinforcement learning is *the equivalent of teaching the machine to play the game.* The machine is provided with a set of allowed actions, rules, and potential end states. The outcome of any single game depends on the *judgment of the player who must adjust his approach in response to the skill and actions of the opponent.*

By applying the rules, exploring different actions, and observing resulting reactions, the machine learns to exploit the rules to create a desired outcome. Thus, determining what series of actions, in what circumstances, will lead to an optimal result.

The current ultimate in AI computer game playing systems is the series designed by Deep Mind Technologies, now owned by Google. *AlphaGo was their first, and plays the board game Go, a game much more complicated than chess.*

In October 2015, AlphaGo defeated the European Go champion, Fan Hui, five games to zero.[228] This was the first time a computer Go program had beaten a professional human player on a full-sized board without handicap.

Subsequent versions of AlphaGo became increasingly powerful. The latest version known as AlphaZero, is completely self-taught. without any learning from "watching" human games.

AlphaZero was able to beat the former version of itself, AlphaGo, at the game of Go *after only eight-hours of self-training.* It won sixty games and lost forty.

Big Data Analytics

Most of us have never heard of an area of AI called Big Data Analytics, (BDA) but we should understand that it is a real phenomenon that is affecting almost every aspect of our lives, including who is governing us, whether our favorite professional sports team is winning, and which products we buy.

The first thing you need to know is that many companies that we interact with, by using their services or buying their products, are collecting large amounts of personal data about us. Then, without our knowledge, they sell it to data brokers.

The *largest data broker* is an international company named *Acxiom*, headquartered in Arkansas. This firm says that it has, on average, *1,500 pieces of information per person on more than 200 million Americans.* That is the Big Data[XCVII] part.

The people who want to influence us, buy these data bases, and use sophisticated, computer-based analytics to manipulate and compile the data into unique profiles of each of us.

These profiles describe who we are, including our age, address, contact information, political affiliation, what we like, what we do not like, what we do, how we do it, etc. *Then, using this profile, they can target us with specific messages that they want us to hear.*

For example, the U.S. Presidential election of 2012 was one of the first to use Big Data Analytics to impact the outcome of the election. The *traditional manner* of segmenting voters for political advertising is to define blocks of people composed of such wide categories as age, gender, race, etc.

Advertising with this level of gross segmentation has to be broad based, "shot gunning" the population in hopes that the right people would see the right message. Some wags call this approach, "spray and pray".

While this classic approach to advertising was used by most of the 2012 candidates for various offices, the emphasis in the marketing

---

[XCVII] **Big data** consists of extremely large data sets that are unstructured and typically relating to human behavior and interactions.

strategies of some campaigns shifted to the use of Big Data Analytics in order to make the difference in the election.

With the more sophisticated, Big Data Analytics approach, the electorate is seen as *a collection of individuals.* Each person can be measured, one-at-a-time, and then *targeted* for a customized contact. Direct mail, e-mail, telephone, door-to-door, and social media channels were all used in 2012 to bring a targeted message to the right person.

The goals of political operatives are *to influence you to, first, get out to vote,* and then to have you *vote for the candidate they support.*

In the 2012 election, it was the Obama campaign that was able to implement this approach much more effectively than the Romney staff. *As a result, Big Data Analytics played a major role in the outcome of that election.*

The Republican National Committee learned their lesson from their 2012 experience, and developed a large, sophisticated, Big Data Analytics operation to use in future elections.

In the 2016 presidential election, the messages from both parties were individually targeted. The people contacting you had scripts designed to push your particular buttons.

For example, if you were a Bernie supporter in the primaries, the script for Hillary pointed out to you how she had adopted a number of Bernie's policies and that Hillary would, if elected President, govern much as he would have.

For Republicans who were members of the NRA, the caller reminded them of Hillary's strong, anti-gun policies, and how Trump promised to defend the 2nd Amendment.

Hillary, as President, they said, will nominate Justices to the Supreme Court who will vote to overturn the Second Amendment, while Trump will appoint Judges who would stand behind this key provision.

Big Data Analytics is also playing *a major role in professional sports.* Most people have heard of "Moneyball". It was a reality-based book and movie about the use of Big Data Analytics in Major League Baseball.

When Billy Beane was the General Manager of the Oakland Athletics in 2002, he was given the job of fielding a winning team while cutting the overall payroll of the players. He turned to Big Data Analytics to evaluate players throughout the league.

Billy Beane

First, Beane used *player data analytics to identify the characteristics of a player in a specific position that were **most important for that player to contribute to wins.*** Beane was then able to use that information to find and recruit *undervalued* players with the matching strengths.

He put together a team of unknowns with a relatively low, total player payroll. Beane's team went on to be competitive with such opponents as the New York Yankees, who had a player payroll twice that of the Athletics. *And the Athletics made it to the playoffs that season for the first time in many years.*

Since then, the use of Big Data Analytics has exploded in many professional sports around the world. One of the largest Big Data Analytics companies serving the professional sports market is *Prozone*, based in Chicago. They cover a range of sports including soccer, football, ice hockey, basketball, and baseball. Worldwide they have over 850 teams as clients.

Major consumer products and services companies are also finding that Big Data Analytics can play a large role in their accomplishments. It is obvious that a key to success for retail companies is keeping their customers happy. And the best way to do that is to provide exactly the products they are looking for, at the prices they want to pay, and then let the consumer know this information. *This is a win-win approach for the company and for the consumer.*

Macy's, the retail department store, is a good example of companies that are using this new technology to enhance their business. Macy's analysts track the social media, like Twitter and Facebook, and use the data to do detailed analyses of what their customers are saying about Macy's products and services. The idea is *to better* **understand the positive and negative sentiments that the** *customers express about Macy's and its goods.*

For example, through *sentiment analysis* the Macy's forecasters may find that people who are sharing tweets about "suits" are also making use of the terms "Alfani" and "Kenneth Cole" frequently.

This information helps the retailer to identify the brands of suits which should be the focus of their marketing campaigns. And then they can inform the customers who they know will be interested.

Big Data Analytics will only grow and become more sophisticated in the future. The amount of data being collected on individuals is increasing and becoming more detailed, and the sophisticated analytics are evolving to better micro target the population. So, we can expect to see an ever-growing impact on our lives from this approach.

Deep Learning

Deep Learning is a more sophisticated version of Machine Learning and is used extensively to analyze Big Data. It is *based on Neural Networks of algorithms,*[XCVIII] and depends on the availability of enormously powerful, computing systems and the Big Data bases.

The term, *Deep Learning,* was introduced to the Machine Learning community by Rina Dechter of University of California,

---

[XCVIII] **Neural network**: is a set of algorithms that are used to analyze Big Data. Their neural name and structure are inspired by the human brain, mimicking the way that biological neurons signal and interact with one another.

Irvine, in 1986.[229] It is the foundation of many *advanced Machine Learning systems* today.

Perhaps most importantly, Deep Learning has vastly improved our ability to **analyze and understand image, sound, and video objects.**

Professor Rina Dechter

The adjective "Deep" in Deep Learning refers to *the* **use of multiple layers** *in the Neural Network. The multiple layers progressively extract higher-level features from the raw input* (see Figure below).

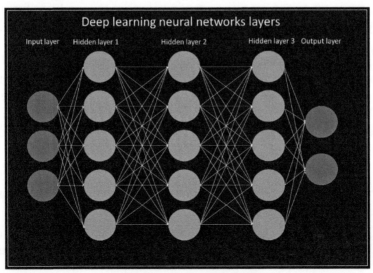

Deep Learning Network

When used in image analysis, with each layer, the Neural Network increases in its complexity, identifying greater portions of the image. Earlier layers focus on simple features, such as colors and edges. As the image data progresses through the layers of the network, it starts to recognize larger elements or shapes of the object *until it finally identifies the intended target.*

Deep Learning is also used with Computer Vision systems as a more sophisticated Machine Learning technique *to better distinguish and classify properties in the images.*

Deep learning algorithms perform a task repeatedly and gradually improve the outcome through the deep layers that enable progressive learning.

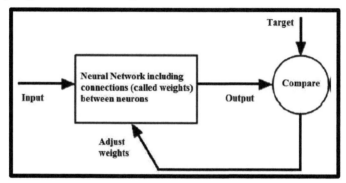

Training process using a Neural Network

The *Deep Learning process to analyze and recognize images with a Neural Network is shown in the flow chart above.* (1) A piece of information about a property of the object being analyzed is **input** to the Neural Network. (2) The data is analyzed in the Neural Network and (3) the **output** is compared to the target image. (4) A change in the property that is determined to better match the target image results in *the weights in the Neural Network being adjusted.* (5) The process continues until the neural network has made enough adjustments (weight changes) to accurately identify the object.

A real-world example of how Deep Learning is used to analyze massive amounts of data is in professional sports. For example, Prozone, the "Moneyball" company discussed in the previous section, places eight video cameras around a soccer stadium to monitor every player and their interactions with every other player. *This system tracks 10 data points per second per player.* *What is then done with that enormous amount of data is amazing.*

Deep Learning algorithms analyze the data in real time and continuously provide the player performance information to the coaches as the game is underway. Coaches are then able to *select the best players for that day* and make *more informed decisions about the strategy and tactics to win the game.*

Yet another example of how this process works is the use of Deep Learning in a Computer Vision system to help screen passengers' luggage before they board their flights. The task can be done much faster and more accurately than depending entirely upon human inspectors.

To *train the passenger screening system*, it was first loaded with a substantial series of luggage images. Some images contained *prohibited items*, such as aerosols, weapons, and liquid containers. Others contained *permitted items*.

Each image was tagged with metadata *indicating the correct answers, either prohibited or permitted. So, the computer had this information with which to work.*

The Deep Learning Neural Network of algorithms processed the visual data, using pattern recognition to identify the many different components of the baggage image. Its outputs, as to whether an item is allowed or not (from the metadata), are fed back into the system, allowing it to learn and improve in accuracy.

Using this technology it is now possible to screen passenger baggage to automatically detect and identify prohibited objects including explosives. The system's throughput rate can be up to 500 bags/hour, which dramatically cuts the time to security check the baggage.

Natural Language Processing

Natural Language Processing (NLP) refers to the branch of artificial intelligence concerned with giving computers the *ability to understand text and spoken words in much the same way as humans.* Its capabilities have reached a high level of utility and now finds wide applications.

NLP combines Linguistics, the scientific study of language, with various computer-based technologies, such as statistical and rule-based modeling as well as Machine Learning. Together, these technologies enable computers to process human language and to *'understand' its full meaning, complete with the speaker or writer's intent and sentiment.*

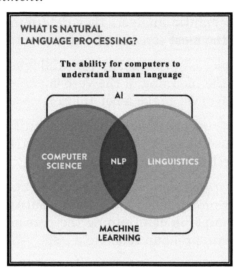

According to IBM,[230] several *NLP tasks* break down human text and voice data in ways that help the computer make sense of the information it is receiving. Some of these tasks include the following:

> *Speech recognition,* also called speech-to-text, is the task of reliably converting voice data into text data. Speech recognition is required for any application that follows voice commands or answers spoken questions like Siri and Alexa.

> What makes speech recognition especially challenging for NLP applications is the way people talk—quickly, slurring words together, with varying emphasis and intonation, in different accents, and often using incorrect grammar.

> *Part of Speech tagging,* also called grammatical tagging, is the process of determining the part of speech of a

particular word, or piece of text, based on its use and context. For example, *Part of Speech* identifies 'make' as a verb in 'I can make a paper plane,' and as a noun in 'What make of car do you own?'

Word sense disambiguation is the selection of the meaning of a word with multiple meanings through a process of semantic analysis that determines the word that makes the most sense in the given context.

For example, word sense disambiguation helps distinguish the meaning of the verb 'make' in 'make the grade' (achieve) vs. 'make a bet' (place).

Named entity recognition, or NEM, identifies words or phrases as useful entities. NEM identifies 'Kentucky' as a location or 'Fred' as a man's name.

Co-reference resolution is the task of identifying if and when two words refer to the same entity. The most common example is determining the person or object to which a certain pronoun refers (e.g., 'she' = 'Mary')

Sentiment analysis extracts from the text such subjective qualities as an individual's attitudes, emotions, sarcasm, confusion or suspicion.

*Natural language **generation*** is sometimes described as the opposite of speech recognition or speech-to-text; *it is the task of putting structured information into human language.*

Natural Language Processing is used extensively in many Machine Intelligence applications. *Here are some examples:*[231]

Spam detection: the best spam detection technologies use NLP's text classification capabilities to scan emails for language that often indicates spam or phishing.

Machine translation: Google Translate is an example of widely available NLP technology at work. *Useful machine translation involves more than replacing words in one language with words of another. Effective translation has to capture accurately the meaning and tone of the input language and translate it to text with the same meaning and desired impact in the output language.*

Virtual assistants and chatbots: Virtual assistants such as Apple's Siri and Amazon's Alexa use speech recognition *to recognize patterns* in voice commands and natural language generation to respond with appropriate action or helpful comments.

Chatbots perform the same function *in response to typed text entries.   A Typical application is* online "help" functions. The best of these also learn to recognize contextual clues about human requests and use them to provide even better responses or options over time.

Social media sentiment analysis: NLP has become an essential business tool for uncovering hidden data insights from social media channels. Sentiment analysis can analyze language used in social media posts, responses, and reviews to extract attitudes and emotions in response to products, promotions, and events. This is information that companies can use in their product designs as well as their advertising and promotion campaigns.

Text summarization: Text summarization uses NLP techniques to digest huge volumes of digital text and create summaries and synopses for indexes and research databases. The best text summarization applications use semantic reasoning and natural language generation (NLG) to add useful context and conclusions to summaries.

The next big leap in NLP capability and utility will come with increased computer power.   The idea of applying quantum computers to NLP first appeared in a research paper by Will Zeng and Bob Coecke[232] in 2016. The authors proposed to incorporate grammatical structure into algorithms that compute meaning. They showed that, while this algorithm has many advantages, its implementation was hampered by the large classical computational resources that it required. But they pointed out that *the computational shortcomings of their approach could be resolved using quantum computation.*

A team of researchers, including Bob Coecke, at Cambridge Quantum Computing, headquartered in Cambridge, UK, say that *quantum computers,* even today's versions, *are significantly better at reading and understanding the meaning of language than their classical counterparts.* The vast improvement in the speed of calculations, has allowed, for the first time, a Cambridge NLP algorithm to be *"meaning aware". This implies that computers can actually understand whole sentences and not just individual words.*

With further development, quantum computer awareness can be expanded to whole paragraphs, and, ultimately, *real time speech. Full scale implementation is dependent on quantum computers becoming much larger in terms of the number of qubits, say 100, than is currently the case.*

Waiting for the Singularity

Most people have not heard of "The Singularity". The concept of the Singularity was first proposed by the Princeton mathematician, *John von Neumann,* in the 1950s. He described *the Singularity as an "event horizon", the moment beyond which "technological progress will become incomprehensively rapid, complicated," unknowable and irreversible."*

At that point in time, technology will be at a level where humans will achieve some unbelievably good things. *At the same time, however, they could lose control resulting in significant unknown, negative consequences.*

Professor John von Neumann

When von Neumann, made his observation (the 1950s), it was difficult to visualize what technological development scenario could lead to such an event.

Now, however, that future is quite clear to many scientists. Artificial Intelligence (AI), has been improving at an *exponential rate* of increase over the last thirty years, and will continue to do so over the next thirty, to the point where some scientists believe von Neumann's prediction about the Singularity could come true before the middle of the 21st century.[XCIX]

[XCIX] See the **Introduction** to this section for the discussion of Moore's Law and the Law of Accelerating Returns.

It will happen when Artificial Superintelligence (ASI) is achieved and there will be an entity on Earth that is, literally, billions of times more intelligent than any human. Shockingly, this could happen in the lifetime of most readers of this book.

Good science fiction writers study the current science related to their stories and build their tales from that base of reality. So, science fiction, in the form of books and movies, may show us what the impact of superintelligent AI might be on our lives.

As discussed previously, the hit movie, **Ex Machina,** portrays a female robot, a gynoid, who was able to use her human-like mind to deceive the people who created her and accomplish her own goal, to be free and out in the world on her own. In the end it cost her creators their lives.

The Arnold Schwarzenegger **Terminator** movies revealed what happens when a net of autonomous, lethal machines reaches a level of intelligence that they become self-aware, turn against their creators, and try to eliminate all humans and "take over the world".

Yet, another popular movie in this genre was **2001: A Space Odyssey.** Hal, the control computer of a spaceship heading for Jupiter, tries to kill all the humans on board its spacecraft because it fears for its life at the hands of the humans.

We are not yet at the level of technological development of Artificial Intelligence where the "reality" shown in these movies is possible. We are getting closer, however, and current AI research is heading in that direction. And, yes, it is an exciting, but, at the same time, a scary prospect.

Many scientists and high-level executives in the technology industries have warned about the dangers of ASI. One of the latest is *Sandar Pichai, Chief Executive Officer of Google.*

Recently, Mr. Pichai gave an interview to the Washington Post in which he said that artificial intelligence holds great promise for the benefit of humanity. But, he said, *concerns by some scientists about the potential for harmful applications of the technology are "very legitimate".*

He mentioned as one example, the development of autonomous AI weapons that can make "kill decisions" on their own. Think about Skynet in the Terminator series of movies. Or the slaughterbots video that shows the danger of that capability.

Google is a leader in the development and application of artificial intelligence. One of their companies, *Deep Mind*, for example, has *developed algorithms that can learn to solve any complex problem without needing to be taught how.*

They have achieved a turning point in developing human-like Artificial Intelligence or Artificial General Intelligence. *Deep Mind's* program called AlphaZero is showing human-like qualities of intuition and creativity. It was made with a "tabula rasa" or a clean slate of knowledge, but with the ability to learn and then remember what it absorbs.

For example, the developers allowed AlphaZero to learn to play chess. Unlike past computers that were programed by the developers to play the game by doing calculations of every possible move, Alpha Zero knew nothing about the game, except the basic rules.

To learn, it played forty-four million matches with itself in nine hours and learned from each one. It got to the point in that short period of time where it was able to beat Grand Masters, but with approaches that were never before employed by a chess computer. It was using human-like intuition and creativity.

Garry Kasparov, former World Chess Champion, said:

"Instead of processing human instructions and knowledge at tremendous speed, as did all previous chess machines, AlphaZero generates its own knowledge. It plays with a very dynamic style, much like my own. The implications go beyond my beloved chessboard."

Deep Mind went on to allow another related program called AlphaGo to learn to play a game even more complex than Chess. That is the ancient Chinese board game of Go. The self-taught machine reached the point in its capabilities where it was able to beat the human, Go champion, Lee Sedol, four games out of five.

We have discussed drones, both military and civilian types. Currently, most of these intricate machines have human pilots on the ground, controlling the craft remotely. But fully autonomous drones have been developed and produced by a number of companies like Amazon. And, ominously, the military organizations of many countries.

Like driverless cars, these drones need detailed sensor input, but in three dimensions, and a sophisticated, onboard AI computer to analyze the data and direct the craft safely to its destination.

Amazon, for example, is well underway to achieving the use of autonomous drones that will deliver small packages to customers' doorsteps in thirty minutes or less after ordering.

The Federal Aviation Administration (FAA) has issued Amazon permission to begin conducting delivery drone operations. An FAA spokesperson said that the Administration issued Amazon Prime Air its certificate on August 29, 2020. The Part 135, air carrier certificate allows Amazon to use "unmanned aircraft systems," or UAS, in a commercial operation.[233]

Amazon Drone Delivery

The capabilities of intelligent machines will only increase as computers and their software continue on their exponential trajectory of improvements. The most sophisticated of these new machines are currently capable of learning with Neural Networks, similar to the way that humans come to understand. But they are faster and more accurate than people.

A few years ago, Harvard University was awarded a $28 million contract by the Intelligence Advanced Projects Activity (IARPA) to study the reasons that human brains are so much more powerful than AI is today. The goal was to ultimately make AI systems faster, smarter and act more like human brains.[234]

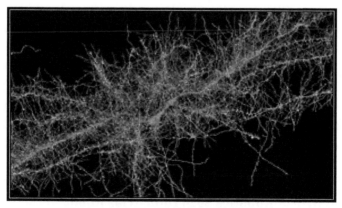

Axons Sent out from a Neuron

The project combines research in brain science with computer science. A detailed, neural map of the brain is being generated. The researchers will then work on understanding better how connections between neurons (as seen in the above Figure) allow the brain to process information.[235] The result, they hope, will be new algorithms that will vastly outperform current AI capabilities.

As AI machines approach even more human-like capabilities, their impact on the world will be deeper and more widespread. Robots have already played a significant role in automation, displacing many workers. The advent of industrial robots started in the '80s. Now, there are close to 300,000 industrial robots in the United States.[c] But that is just scratching the surface.

As robots combine human-like agility, as demonstrated in Boston Dynamics Atlas robot, with increased intelligence, improved vision, Natural Language capabilities, and improved Deep Learning, the more they will be able to take over for humans in complex manufacturing and service tasks. *And, as a result, more people will lose their jobs.*

Dealing with this unemployment issue will be one of the major collateral issues that humankind will have to face as we approach the Singularity.

For example, as previously noted, Level 5 self-driving cars and trucks will be common in the next twenty-five years. Uber and taxi drivers will be replaced. Driverless, long-haul trucks will make dock-to-dock, cross country runs faster and cheaper than with humans.

Perhaps of even more concern to the future of mankind than the loss of jobs is the use of lethal military, Artificial Intelligence systems that incorporate advanced learning software. Military organizations around the world are creating a range of *Lethal Autonomous Weapons* (LAWs). They include mobile, track-

---

[c] The new **World Robotics 2020 Industrial Robots** report presented by the International Federation of Robotics (IFR) shows a record of about 293,000 industrial robots operating in factories of the United States – an increase of 7% over the previous year.

propelled, tank-like machines mounted with heavy fire power; and jet-propelled, combat drone aircraft that can autonomously search, identify, locate, and destroy enemies. Other autonomous weapon systems are being developed to work on and under the sea, as well as in space.

The military of other countries, especially Russia and China, are following suit. Some fear that it could lead to a new AI arms race.

All of these developments have spawned an ever-widening level of concern among some scientists and ethicists. *Will humans lose control of the Artificial Intelligence?*

Nick Bostrom is a Swedish-born philosopher at the University of Oxford. He is the founding director of the Future of Humanity Institute at Oxford University. One of his areas of focus is the dangers of Superintelligence and the strategies for dealing with it.[236]

Nick Bostrom

Bostrom noted that when one compares machine-based neurons to biological neurons, the machine components are already significantly faster. Bostrom writes,

"Biological neurons operate at a peak speed of about 200 Hz, a full seven orders of magnitude slower than a modern microprocessor (about 2 GHz).

Moreover, neurons transmit spike signals across axons at no greater than 120 m/s, whereas existing electronic processing cores can communicate optically at the speed of light.

Thus, the simplest example of a superintelligence may be an emulated human mind run on much faster hardware than the brain. A human-like reasoner that could think millions of times faster than current humans would have a dominant advantage in most reasoning tasks, particularly ones that require haste or long strings of actions."

What Bostrom is saying is that AI that just reaches the level of intelligence of a human, would, *by virtue of its much higher speed of processing, be considered a superintelligence.*

The famous British astrophysicist, Stephen Hawking, was afraid of the advent of artificial superintelligence. Just before he died, he speculated that:

"It (AI) would take off on its own, and re-design itself at an ever-increasing rate. Humans, who are limited by slow biological evolution, could not compete, and would soon be superseded. The development of full, artificial intelligence could spell the end of the human race,"

concluded Hawking.[237] Other scientists agree, but not all.

Ray Kurzweil, Google's Director of Engineering, is a former MIT professor, inventor, famous futurist, and entrepreneur. He is heading up a team that is developing machine intelligence and natural language understanding.

Kurzweil has done a great deal of thinking about the coming Singularity. His track record is about 86% for having his many technological predictions come true.

"Of the 147 predictions that Kurzweil has made since the 1990's, fully 115 of them have turned out to be correct, and another 12 have turned out to be "essentially correct" (off

by a year or two), giving his predictions a stunning 86% accuracy rate." [CI]

Kurzweil sees it differently than Hawking. He recognizes the potential dangers of AI but believes that there are strategies we can deploy to keep AI safe for humans. He said,

"There are efforts at universities and companies to develop AI safety strategies and guidelines, some of which are already in place."

Kurzweil's bottom line is that we have a moral imperative to realize the promise of AI, while controlling its peril. "It won't be the first time we've succeeded in doing such a thing."

What he is talking about is the two-edged sword that many technologies represent. Fire can cook your food or burn our house down. Airplanes can carry passengers or bombs. Rockets can be used to launch satellites or deliver weapons of mass destruction.

Kurzweil believes that the next major improvement in computer technology will come from the brain-computer study that is currently underway at Harvard, to understand in fine detail, how the human brain works.

By the mid-2020's, he says, we will use that knowledge to build an effective model of human intelligence and apply it to computers. *Kurzweil then projects those computers, and their AI software, will be **at the level of human intelligence by 2029. This is termed Artificial General Intelligence (AGI).***

Philosopher David Chalmers argues that AGI is a highly likely path to superhuman intelligence. Once AI achieves equivalence to human intelligence, it will quickly self-extend to surpass human intelligence, and will continue until it completely dominates humans.[238]

Ray Kurzweil agrees. Human-level intelligent machines will be followed by continued exponential growth in computing capacity and capability. By the early 2030s, Kurzweil says, the amount of non-biological computation will exceed the "capacity of all living biological human intelligence." He goes on,

---

[CI] Peter Diamandis in https://www.diamandis.com

"I set the date for the Singularity—representing a profound and disruptive transformation in human capability—as 2045.

*"This is when AI will reach the Superintelligence level, and where it will be difficult or impossible for present-day humans to predict the impact on our civilization. **It will be the Singularity.**"*

At this point, the resulting Superintelligence will be able to develop its own "next-generations" and do these developments at an increasingly higher rate of speed. This "runaway reaction of self-improvement, resulting in an intelligence explosion", as I.J. Good said in 1965, **is where humans could lose all control, and become less important to the AI.**

Machines will become superintelligent, but will they become conscious? Scientists do not know with certainty, but many believe that machines will become conscious and self-aware when they achieve a critical level of complexity.

Neurons, the basic building blocks of human intelligence, are not conscious entities, but, put eighty-six billion of them into a human brain and you have a conscious being.

When the Superintelligence emerges, it will not depend on human programmers. *It will grow to super levels of intelligence through its own programming.* And it will be far more efficient and effective than an army of human programmers could have ever produced.

Somewhere along the way, and we have no idea when or how, a form of consciousness could emerge in the Superintelligence. It may not be exactly like our type of consciousness, but it will likely have similar properties including self-awareness and self-preservation.

Living with the Superintelligence

Many people today are genuinely concerned about Climate Change and the impact it will have on the planet over the next eighty to one hundred years. There is, however, another threat to humanity that is much more alarming, very real, and remarkably close. That is *the advent of Artificial Superintelligence (ASI).*

A superintelligent AI system, not purposefully built to respect our values, could take actions in pursuit of its tasks that neglect our needs. It is not an overstatement to say *that the result could lead to a global catastrophe or even human extinction.*

Building a Superintelligent AI could be like acquiring the King Midas touch. It would be great to have the power to turn anything you touch into gold. But if you cannot turn the power off, you soon encounter a disastrous predicament. Your food and drink also turn into gold. As does anyone you touch. So, rather than a boon, the "Midas touch" turns out to be deadly for its owner and his loved ones.

There are many situations where an out-of-control ASI could turn into a "Midas" problem for humans. Here is one scenario.

Let us say that the Superintelligence is given a goal to maximize the production of paper clips (or any other product for that matter). [239] Without constraints, the ASI will do everything in its power to carry out its mission. It will need vast amounts of raw materials and the corresponding manufacturing facilities to maximize the production of the paper clips.

Yet, we might prefer to use those material resources and production capabilities in other ways. But the Superintelligent system would be amazingly effective at acquiring them in spite of our wishes.

So, the Super Intelligence ends up with an optimized paper clip manufacturing system, to the exclusion of everything else. It essentially consumes the world and its resources, and it has a vast supply of useless paperclips.

This begs the question, *"How do we ensure that the ASI will do what we really want, and not terminate all of us to accomplish its goals?"* What *prior precautions* can the programmers take to successfully prevent the Superintelligence from catastrophically misbehaving once it emerges?

Many people have been working on strategies to avoid these challenges. One class of control is *to limit the ASI's ability to influence the world.* A second is *to control its motivations, and to build into the ASI goals that are congruent with human values.*[240]

But it is important to understand that *any attempts to solve the control problem **after** the superintelligence is created, would fail.* The reason is that a *Superintelligence would be vastly superior to all humans.* And it would be dominant over us, just we are dominant over animals. With t*he AI's decisive strategic advantage, it would be impossible to constrain or limit it in any way.*

There is still a great deal of work to be done to find an approach that will work. Keep in mind that we will only get one opportunity to implement a solution. If it does not work the first time, *we will not get a second chance.* The Superintelligence will not let us.

Nick Bostrom of Oxford University has considered some of the possibilities.[241] In his book, **Superintelligence**, he analyzes the two broad classes of potential methods for removing the threat of the Superintelligence. They are *capability control and motivation selection.* There are several specific techniques within each class. *What follows is a discussion of some his ideas.*

Capability Control

Capability control methods seek to prevent undesirable outcomes by **limiting what the Superintelligence can do.** There are several approaches that have been considered:

1. Boxing in
2. Socializing and training
3. Stunting
4. Using Tripwires

Boxing in – This is an approach that attempts to *confine the ASI in its abilities to perform.* That is, to "box it in". The system would have a set of protocols that contains and constrains the ways in which the Superintelligence can interact with the world.

For example, if we limit the kinds of information it receives, and minimize the number of actions it can perform, we will have a way to control the threat it might represent.

There are a number of problems with this approach. First, if humans are in the control loop, they are *vulnerable to manipulation by the Superintelligence.*

A superintelligent AI could trick its human gatekeepers into letting it out of the box. Remember the movie Ex Machina, this is exactly what happened to the humans.

Second, the Superintelligence *could find a creative way out of its box.* And it could succeed in spite of all of our best thinking beforehand.

Third, limiting the ASI's ability to act may indeed keep it confined, but it also limits its ability to do what we want it to do. The owners/creators of the ASI, therefore, may not accept this approach as a solution to the problem, since they will not achieve what they want from the ASI.

Socializing and Training - A second containment approach involves socializing and training the ASI to live within an *incentivized environment.* In this approach, the ASI receives *some significant benefit* by cooperating with humans, and that forces the ASI to behave in morally and socially acceptable ways. *The hope is that, by living in this environment,* **the ASI will acquire human norms and values.**

Bostrom sees some issues with the approach. *The major problem is that the reward/benefit that the Superintelligence receives* **would be administered by humans.** Therefore, the *benefit would be subject to unknown possibilities of loss by the ASI, and* **the Superintelligence would recognize that fact.**

For example, the ASI may come to *distrust the humans* who administer the benefit. It cannot be sure that the humans *will be able* to provide the "promised" reward.

For example, *the ASI may "worry"* about the *humans becoming incapacitated.* Or the possibility of a "regime change" where the cooperating humans are replaced. And their replacements, for some reason, would not follow through with the "agreement".

All of these "thoughts" on the part of the ASI would lead it to use its superior intellect to *find a means to ensure that none of these problems occur. And any one of its "solutions" for these human caused problems may be detrimental to those humans.*

Stunting – by this, Bostrom means that *the creators somehow hamper or disable the ASI.* One approach is to make the Superintelligence *operate on slow hardware.* Another possibility would be to *reduce its memory capacity or limit the kind of data it can process.*

The problem with this approach is getting the balance just right. If you stunt it too much, then it will not be superintelligent. If you do not apply enough constraints, then it will quickly figure out how to overcome its barriers.

Using Tripwires – involves building into the ASI system a set of behavioral or ability "tripwires" which, if crossed, *shuts down the project and destroys it.* Some examples are behavioral and ability tripwires.

A *behavioral tripwire* consists of a network of detectors placed around a boxed-in AI. They would detect any attempt to breach the containment.

An *ability tripwire* would check to find if any new, unexpected abilities are found in the AI. If the new ability is thought to be leading to a "dangerous" skill, the Superintelligence would be shut down.

Like the other methodologies, *tripwires could be vulnerable to defeat by the ASI.* With its superior intellect, the ASI could study and identify ways and means to subvert a tripwire in an unexpected or unanticipated way.

Motivation Selection

Bostrom covers a *second class of ASI controls* designed to stop the Superintelligence from becoming a threat to humanity. They are called "motivation selection" techniques. *The idea with these approaches is to **ensure that the ASI has benevolent or non-threatening motivations built into it, and that its interests are congruent with those of its human creators.***

These motivation selection approaches include:

336

1. Direct Specification

2. Domestication

3. Indirect Normativity

Direct Specification – with this approach, the *creators program the ASI with the desired set of motivations.* A good example of this approach is Isaac Asimov's *"three laws of robotics"*, as outlined in his fictional books and short stories about living with robots.

The laws are:

1. A robot may not injure a human being or, by failing to act, allow a human being to come to harm.

2. A robot must obey orders given to it by human beings, except where carrying out those orders would break the First Law.

3. A robot must protect its own existence, as long as the things it does to protect itself do not break the First or Second Law.

It would seem at first glance that using these three rules would be a "foolproof" method to control the ASI. However, Asimov spent much of his stories describing how the three rules are vague and subject to interpretation. And that t*hey could be bent to create harmful situations for humans.*[242]

One example of Asimov's point about the problems with the "laws" is in *the interpretation of the First Law.* It says that *a robot may not, through inaction, allow any human to come to harm.*

This implies that the Superintelligence must be constantly watching to *avoid* all possible ways in which humans might be hurt. But humans, throughout their lives, are always getting injured in many different ways. So, how can the ASI stop all such things from happening?

A Superintelligence with control over humans might decide that the safest thing to do would be to put all of its humans into artificially induced comas. It would not be a great life for them, but it would certainly prevent them from coming to any harm during an active life.

Domestication – just as wild animals were bred and trained to be used domestically, so too ASI could be "domesticated". Domesticated animals lack the drive or motivation to do anything that might harm their human owners. They are happy to work within the domestic environment and their behavior can be controlled in that environment through the use of rewards and incentives.

Likewise, the ASI would be motivated to work and act on a small scale, within a narrow context and through a limited set of action modes.[243]

This approach, however, severely limits the ASI's functionality. Would it be worth it to have an ASI if it cannot operate as a true Superintelligence?

Moreover, when implementing this method, one has to be concerned with exactly how to align the wants and needs of the ASI with those of the creators. This presents a significant engineering challenge to reach the *optimum solution. Any failure to properly align wants and needs could be disastrous for the humans.*

Indirect normativity – is Bostrom's third motivational approach to taming an ASI. The idea here is to provide the ASI *with some procedure or ability to determine its own ethical and moral standards.* **The AI is programmed to function like an ideal, hyper-rational human being**. It would, if all went well, produce what humans would have wished if they had pondered the question long and hard enough. [244]

A major problem with this approach, obviously, is *ensuring that you have the right "norm-selection" procedure. Getting it slightly wrong could have devastating implications,* particularly if the machine has a decisive strategic advantage over us.

Summary Conclusion for all containment approaches - Bostrom, in his general analysis of the range of containment approaches, argues that *it is virtually impossible to have a perfectly isolated system. Any interaction with a human agent would compromise the isolation and doom the containment.*

So, given whatever solution is adopted, things can still go wrong. ASI systems are not perfect, and there may be unintended consequences from any given specification. Therefore, emergent behavior of the ASI can diverge dramatically from the design intentions that were built into the controls.

Bostrom's bottom line is that implementing any one of the approaches is exceedingly difficult. A Superintelligence could probably overcome them all.

So, the search for a certain solution must continue, and be in place before the Singularity. Otherwise, it will be too late.

Direct Human Link to the ASI

Elon Musk, the creator of Tesla electric cars and SpaceX rockets, was interviewed for a recent documentary by American filmmaker Chris Paine, called "Do you trust your computer?"

Musk warned that *Artificial Superintelligence can leave humanity behind and lead to the creation of an "immortal dictator" who will control the world.*

> *"At least when there's an evil dictator (that is human), that human is going to die. But for an ASI (based dictator), there would be no death. It would live forever. And then we'd have an immortal dictator from which we can never escape."*

Musk said that *ASI does not have to be evil to be a problem for humanity.*

> *"If ASI has a goal and humanity just happens to be in the way, it will destroy humanity as a matter of course without even thinking about it. No hard feelings."*

Elon Musk and Ray Kurzweil agree on one important thing. *Humans must, somehow, merge with the computer to stay relevant in the world of Artificial Superintelligence.*

Musk has formed a San Francisco-based company, *Neuralink*, to achieve that goal. Neuralink is developing *ultra-high bandwidth* (the speed and capacity of the connection), *implantable, brain-machine interfaces to connect humans and computers.*[245]

Here is how Neuralink works. *A small device, about the size of a quarter, is placed in a hole that is cut out in the skull.* There are a network of very thin wires emanating from the mechanism.

These wires are connected to important parts of the brain. The small electric threads are thinner than a human hair and can send signals in both directions. They will receive impulses from neurons and also be able to influence the neurons.

The device's battery is designed to last for a full day and is charged wirelessly in the evening.

As a near-term application for the Neuralink, the company aims to help paralyzed patients interact with their phones or computers using the brain chip. Neuralink has already started testing the chip on animals and recently showed the results of a chip that was placed in a pig.[246]

Although the pig demonstration showed neural activity being broadcast wirelessly to a computer, it did not show that the computer understood what the spikes of neural activity actually meant. Nor did the demonstration show the computer usefully communicating back to the pig's brain.

Musk also said, but did not demonstrate, that a Neuralink chip was implanted in a monkey's brain and the simian was able to play video games using nothing but its mind to control the computer. *They plan to start human trials in late 2021.*

A major issue with the Neuralink technology is that it is so invasive. Many people will be repelled by the idea of having a hole cut in their skull and wires inserted into their brain. Other, less invasive methods would be more attractive.

A group of researchers at the University of Melbourne, in Australia, have leapfrogged Neuralink and placed electrodes *into the brains of humans* using a much less invasive technology. The scientists inserted the electrodes through the jugular vein in the neck and pushed them up into the brain's primary motor cortex. Once there, the electrodes were nestled into the wall of the blood vessel where they could detect brain signals and feed them back to a Windows 10 computer.[247]

As an experiment to test the utility of the methodology, two humans with advanced ALS, a highly degenerative disease, had the electrodes inserted into their brains with this methodology.

After a couple of months training, *the patients were able to control the mouse of a computer by just using their thoughts*.

> "The participants undertook machine-learning-assisted training to use a wirelessly transmitted electrocorticography signal associated with attempted movements to control multiple mouse-click actions, including zoom and left-click,"

The results of the study were published recently in an article in the *Journal of NeuroInterventional Surgery*.

Ray Kurzweil has a similar vision to Elon Musk to resolve the threat posed by ASI. He also foresees a computer-mind connection, but of a different type than those developed by Musk or the University of Melbourne. Kurzweil calls it *a neocortex connection, and it is made using nanobots.*

Nanobots are molecule size devices that are injected into the blood stream to carry out pre-programmed tasks. A nanobot's (or "bot") components are near the scale of a nanometer, or 10^{-9} meters. As a comparative, DNA molecules are about 2.5 nanometers wide.

Kurzweil has predicted that *by the 2030s*, nanobots capable of tapping into our neocortex and connecting us directly to the world around us, *will be available to humans.* His idea is to use the "bots" to connect your brain directly to the internet, thereby upgrading your intelligence and memory capacity by orders of magnitude. Thus, as the machines become smarter, so do we.

He has said that *this ability to expand our brains with the information held in the cloud will combine with the power of artificial intelligence to make humans superintelligent.*

But, *how far along is the development of the nanobots* that could make this connection a reality? The first actual use of nanobots for a medical application was carried out in early 2020.

Researchers at Arizona State University injected nanobots into the bloodstream of mice afflicted with cancer.[248] *The nanobots were made from DNA nanostructures (DNA that is folded in order to measure 90nm) along with the blood-clotting enzyme, thrombin.*

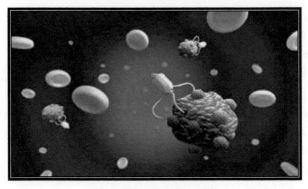

Nanobots at work

The nanobots targeted a protein called nucleolin, which is found only on the surface of cancer cells. After attachment to the cancer cell, the nanobots release thrombin into the cell. This cuts off the blood supply to the cancer cells, thus destroying the tumor.

The nanobots shown in the Figure *above are visual metaphors for the real bots,* which are folded DNA structures in the case mentioned above.

The nanobots work fast and in huge numbers, completely surrounding the tumor. The experiment was successful and drastically shrunk the tumor, leaving the healthy cells unaffected.

According to Kurzweil, over the next ten years, nanobot technology will continue to grow at an exponential rate. It could, by the early 2030's, conceivably be able to accomplish Kurzweil's vision of connecting the human neocortex to the cloud.

Other complementary technologies will have to be developed to supplement the nanobots. For example, DNA already has the potential to transform the computing world by recreating living cells into data storage devices.[249]

Nanorobots, which can function like living computers, are created using DNA strands that fold and unfold like origami. Daniel Levner, a bioengineer at the Wyss Institute at Harvard University, and his colleagues at Bar Ilan University in Ramat-Gan, Israel, made such nanobots by exploiting the binding properties of DNA.[250]

Bioengineers agree with Kurzweil and believe that DNA nanobots could carry out complex programs that will one day be used to connect a human's neocortex to the cloud. This accomplishment will allow people to send e-mails and photos directly to each other's brains while also backing up their thoughts and memories on the cloud.

Thus, instead of humans becoming obsolete, we could be working with the machines. But it is important that humankind perfect this human-machine connection technology before the Singularity and ASI emerges. *If we cross over that event horizon and are not working with the machines, we may not be given the chance later.*

Part IV

Extraterrestrial

Intelligence

IV. Extraterrestrial Intelligence

Introduction

Are we alone in the universe? When we look up at the sky on a dark, clear night and see the myriad of stars across the heavens, we cannot help but ask ourselves this question. With so many stars, we think, *there must be many civilizations like ours out there.*

But it is a lot *more complicated than just the vast number of stars. In fact, as we will see, there is* **a good chance that we are indeed alone!**

NASA now believes that there are more than *100 billion stars in our Milky Way galaxy*. Most of them are in the densely packed, inner bulge of the galaxy, shown as the bright, white blob in the Figure. The rest of the stars are in the spiral arms, or close to them.

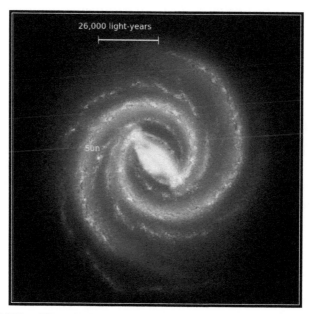

The Milky Way Galaxy, an artist's view seen from above

347

The galaxy is relatively flat when seen from the side. The astronomers call it a barred, spiral galaxy. It has an estimated visible diameter of more than 100,000 light-years.[CII]

As shown, if you look carefully, in the Figure above, our Sun is located in one of the Milky Way's spiral arms, about one third of the way out from the center. Fortunately for us, this is in the *"Galactic Habitable Zone"* for life on planets in the galaxy.

There is not wide agreement on the size of this Galactic Habitable Zone, but *one estimate is that it only consists of about 10% of the total galaxy.* And, because of where it is located, it is *in a region of the galaxy that is star poor.* **So, the bottom line is that there are relatively few stars with exoplanets in the Galactic Habitable Zone that are also life friendly.** *This is one strong, negative indicator for the probability that intelligent life exists elsewhere in our galaxy. There are many others.*

SETI's Estimate of the Number of Habitable Planets

The stars *in our Milky Way galaxy* are what we see as we look up toward the sky. But, according to NASA, there are almost **one trillion other galaxies** *in the visible universe*, each with as many stars and planets as the Milky Way. So, the numbers of planets out there are even more vast than we initially imagined.

The Milky Way got its name because, when we gaze up, it appears as a hazy band of milky light, formed by the billions of tightly packed stars that cannot be individually distinguished by the naked eye.

Over the last thirty years, using some new technologies, like the Kepler telescope, we now know that most of the 100 billion stars in the galaxy have planets around them. *We call them* **exoplanets** *because they are outside of our solar system.*

[CII] **A light year** is the distance that light, *travelling at 186.3 thousand miles per second,* can cover in one year, or 5.9×10^{12} miles. So, 100,000 light years is 5.9×10^{17} miles (about 6 followed by 17 zeros!).

The Milky Way

NASA estimates that the Milky Way has at least *100 billion exoplanets*, averaging about one per star. *But how many of them are candidates for having advanced life? This is an open question being investigated by many astronomers and astrobiologists.*

The **SETI** (Search for Extraterrestrial Intelligence) Institute, a private, non-profit organization in California, is extremely interested in identifying the *locations of stars with planets that are likely to support life.* With this knowledge, they can focus on those stars to search for intelligent signals.

Jeff Coughlin is an exoplanet researcher at the SETI Institute. He was part of the team that produced a recent estimate of the number of exoplanets in the Milky Way that may have developed life.[251]

The astronomers focused on three main determinants for habitability. First, they estimated the number of exoplanets *similar in size to Earth.* These are most likely to be rocky planets like ours.

Then, they looked at *how many stars are similar in age and temperature to our Sun.* Lastly, they considered whether the planets have the conditions necessary to support liquid water, a critical ingredient for life. These *planets must be found in the "Goldilocks' zone" around their star.* Here, it is not too cold and not too hot, and water can exist in liquid form. With these conditions, the SETI scientists believe, life as we know it has a chance to develop.

The SETI astronomers' calculations resulted in an estimate of only 300 *million* of the 100 *billion* exoplanets in the Milky Way as being habitable. It still sounds like a lot, but it is only 0.3% of the total number of planets in the galaxy.

According to other astronomers and astrobiologists from various academic institutions, **even this small number established by SETI is way too large.** To have a good chance for life, those planets also need to be located in *the Galactic Habitable Zone.*

The Galactic Habitable Zone

The galactic regions *that are hospitable to life can be termed "Galactic habitable zones"*. This term was coined by Guillermo Gonzalez, Assistant Professor of Astronomy at the University of Washington.[252]

The Galactic habitable zone is:

- Free of excessive amounts of high energy radiation,

    o Gamma rays[CIII], x rays and cosmic rays

- Not crowded by other stars that would increase the gravitational effects in the region,

- At a location in the galaxy that is relatively close to the center where there are heavy elements, created by supernovae[CIV], and which are needed for life,

    o but far enough away to avoid the excessive amounts of radiation emitted by collapsing stars also found there,

---

[CIII] **A gamma ray**, also known as gamma radiation (symbol γ or ɣ), is a penetrating form of electromagnetic radiation arising from the radioactive decay of atomic nuclei.

[CIV] **Supernova** - a star that suddenly increases greatly in brightness because of a catastrophic explosion that ejects most of its mass.

o   and the high gravity associated with densely packed stars.

o   In addition to the factors above, an exoplanet *needs to have its own magnetic field* to divert cosmic rays from impinging the surface. Otherwise, the cosmic rays would damage any life forms, and ultimately strip away the atmosphere of the planet.

A more detailed discussion of each of these factors follows.

For life to develop on an Earth-like planet that is in the "Goldilocks zone" of its star, the planet *must also be free from the bombardment of excessive amounts of high energy radiation.* This includes the x-rays and gamma rays that are generated in various cosmic events. *If the planet is in a region of the galaxy that has extreme amounts of this type of radiation, e.g. near areas with many supernovae, then life will be difficult to develop.*

Cosmic rays are another type of dangerous radiation. They consist of *high-energy* protons and atomic nuclei that move through space at nearly the speed of light. *Some of these particles are created in the highly energetic reactions that occur in the Sun.*

Others come from outside of the solar system but originate in other stars in our own galaxy. Then, there are even a small number of cosmic rays that reach us from the reactions that occurred "a long time ago in a galaxy far, far away".

In Earth's case, *we are protected from these cosmic rays by our relatively strong magnetic field* which diverts the charged particles away from the surface of Earth. We see the effect when we look at the Northern Lights or Aurora Borealis. In the Southern Hemisphere, the lights are called Aurora Australis.

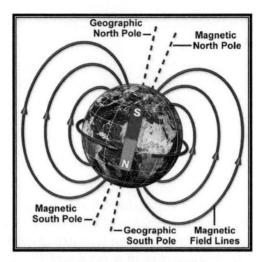

Earth's magnetic field

The Earth's magnetic field is like a bar magnet whose field lines end at Earth's poles as shown above. As the trapped cosmic ray particles[CV] spiral along the field lines, they come into the atmosphere near the poles, and collide with the nitrogen and oxygen atoms in the atmosphere. They then radiate energy in the form of light as a result of the collision. *These are the Auroras.* This occurs between 60 to 600 miles above the Earth's surface.

Unfortunately, not all exoplanets have magnetic fields to provide this protection. *So, having a strong enough magnetic field becomes another requirement for life, as we know it, to develop on an exoplanet.*

Mars, for example, has no magnetic field. Potential life that may have developed there would have been bombarded by the Cosmic rays and been severely damaged. Another consequence of cosmic ray bombardment was that the Red Planet had its atmosphere stripped away billions of years ago by these energetic particles.

So, an exoplanet without a magnetic field is likely to not have an atmosphere. Therefore, no life could evolve into intelligent beings.

[CV] A charged particle moving in a magnetic field is "captured". The motion of the particle will be a spiraling in the direction of the magnetic field. On Earth, the particles spiral into the poles, colliding with atmospheric molecules, creating the Auroras.

Yet another potential negative factor for life to form on an exoplanet is that it *cannot* be in a region of the galaxy that is crowded with close-by stars. *The gravitational effects of all of those bodies can cause problems for the formation of life.*

For example, we know that gravity affects key molecules that control cellular processes like growth. In stronger gravitational fields, the size of cells decreases; and in weaker gravitational fields the size of cells increases. Gravity is, therefore, a limiting factor in the growth of individual cells. *Thus, gravitational effects may prevent conditions from being conducive to life.*

A planet's *relative location in the overall, galactic structure* also affects its ability to develop life forms. To have enough of the heavy elements needed for life, an *exoplanet and its star need to be fairly close to the galactic center* where most of these materials were formed in the supernovae of dying stars.

But since the inner region of the galaxy, near the center, is populated with supermassive stars that are about to die, this creates a complication. When these stars turn into supernovae, they also *produce high amounts of radiation that is harmful to any life in the region.*

The Bottom Line

Given all of these constraints, scientists have identified a "Galactic Habitable Zone" in a spiral galaxy. It is between the spiral arms that are more than a third of the way out, but, not too close to the edge. **This is the position where Earth is found in our galaxy.**

There is not wide agreement on the size of this galactic habitable zone, but *one estimate is that it is only about 10% of the total galaxy.* And, because of where it is located, it is in a region of the galaxy that is star poor. **So, the bottom line is that there are relatively few stars with exoplanets in the galactic habitable zone that are also life friendly.** *This is a strong, negative indicator for the probability that other intelligent life exists elsewhere in the galaxy.*

The Formation of Intelligent Life

Even though an exoplanet passes all of the criteria that makes it compatible with the formation of life, *and* its star is in the Galactic habitable zone, *it still may not develop complex life forms*[CVI] *that could lead to intelligent beings.*

To this day, *it is a mystery why and how life developed on Earth. We need to understand that process first before we can decide if life is possible on an exoplanet. And we are far from achieving that goal.*

Astrobiology[CVII] is the field of investigation that considers how life is formed from non-living materials. More broadly, it is *the study of the origin, evolution, and distribution of life in the universe.* A central goal of the field is to find evidence of past or present life *beyond Earth.*

To accomplish this objective, astrobiologists try to understand how life originates and how it can survive in many diverse types of environments. This often involves the study of extreme life right here on Earth.

One extreme species is the Thermococcus microbe. These organisms were found living near deep-sea, hydrothermal vents near Papua New Guinea. It is in this location where super-hot water, heated by the magma[CVIII], seeps out of the Earth's crust. These microbes can survive on so little energy that, until now, the chemical reaction it uses wasn't thought able to sustain life.

[CVI] **"What is life?"** While there is no consensus among scientists, according to NASA, one frequently used definition is *"a self-sustaining system capable of Darwinian evolution."*

[CVII] **Astrobiology**, formerly known as exobiology, is an interdisciplinary scientific field that studies the origins, early evolution, distribution, and future of life in the universe. Astrobiology considers the question of whether extraterrestrial life exists, and if it does, how humans can detect it.

[CVIII] Magma - hot fluid, or semifluid, material below or within the earth's crust from which lava and other igneous rock is formed on cooling.

The Beginning of Life on Earth

On our planet, the emergence of complex, intelligent life required a series of unlikely, intricate, evolutionary transitions **starting with *abiogenesis, the creation of life from non-living materials*.** Then, the development of multicellularity, and, ultimately, evolution leading to intelligence itself.

So, how did it all begin? Earth was formed about 4.5 billion years ago. We know from the fossil record that life, from abiogenesis, appeared on Earth about one billion years later, soon after the oceans appeared.

There are many theories about how that life developed on its own, *but still no consensus among scientists. Most of these ideas assume that cells are too complex to have formed all at once,* so, they say, life must have started with just one part that survived. Then, somehow, was joined with the other necessary components that, by some means, were found around it.

Chemists at Scripps Research in La Jolla, California, led by Dr. Ramanarayanan Krishnamurthy, *believe they may have found the path to abiogenesis.* They showed in the laboratory that a simple compound called diamidophosphate (DAP), which was plausibly present on Earth before life arose, could have chemically knitted together tiny DNA building blocks, called deoxynucleosides, into strands of primordial DNA (deoxyribonucleic acid).[253]

Research results over the past several years, have pointed to the possibility that DNA, and its close chemical cousin RNA[CIX], arose together as products of similar chemical reactions, and that the first self-replicating molecules, the first life forms on Earth, were mixes of the two.

[CIX] **RNA** (ribonucleic acid) can both serve as a genetic code and catalyze chemical reactions. Its principal role is to function as a messenger carrying instructions from DNA for controlling the synthesis of proteins.
.

Dr. R. Krishnamurthy, Scripps Institute

However, no one has yet been able to produce any type of life form in a laboratory. Such a result would prove the viability of the concept that life started from some particular mix of non-living materials. *So, the actual process is still a mystery*.

Another theory that hypothesizes an explanation has been proposed in the last few years.[254] The idea is that an ancient cataclysm may have jump-started life on Earth.

This scenario suggests that some 4.47 billion years ago—a mere 60 million years after Earth took shape and 40 million years after the moon formed—a moon-size object sideswiped Earth and exploded into an orbiting cloud of molten iron and other debris.

> *"The metallic hailstorm that ensued likely lasted years, if not centuries, ripping oxygen atoms from water molecules and leaving hydrogen behind. The oxygens were then free to link with iron, creating vast, rust-colored deposits of iron oxide across our planet's surface. The hydrogen remaining in the atmosphere formed a dense envelope of gases that lasted 200 million years as it ever so slowly dissipated into space.*
>
> *After things cooled down, simple organic molecules began to form under the blanket of hydrogen. Those molecules, some scientists think, eventually linked up to form RNA, a molecular player long credited as being essential for life's dawn. In short, the stage for life's emergence was set by a random event almost as soon as our planet was born."*

One researcher, Andrej Lupták, a chemist at the University of California (UC), Irvine, commenting on the relevance of this theory, said:

> *"Fifteen years ago, we only had a few hazy ideas about how life may have come about. Now, we're seeing more and more pieces come together."*

Yet, whichever theory proves to be correct, life did develop in the primordial Earth. We know that most organisms, from the first appearance of life forms, were simple ones. They were composed of individual cells, or small multicellular organisms, occasionally organized into colonies.

In fact, for the **next three billion years**, life existed as only these very primitive creatures. *This primordial world lasted, therefore, for 78% of the Earth's existence to date.*

Then, *suddenly, something happened, we do not know what* it was, that *caused complex life to develop* and *explode on Earth*. As the rate of diversification later accelerated, the variety of life became much more complex, and began to resemble that which we see today. *All present-day animal phyla appeared during this period.*

It is called the **Cambrian Explosion**, named for the time period in which it occurred, about 550 million years ago. *It lasted for a truly brief period of geologic time, only about 13 to 25 million years,* **and resulted in the divergence of most of the modern metazoan**[CX] **phyla.**

[CX] **Metazoan**: any of a group (*Metazoa*) that comprises all animals having the body composed of cells differentiated into tissues and organs and usually a digestive cavity lined with specialized cells.

Cambrian Trilobites

The seemingly rapid appearance of fossils in the "Primordial Strata" was first noted by William Buckland, an English paleontologist and geologist, in the 1840s.

The knowledge of the Cambrian Explosion confused Charles Darwin as he developed his Theory of Evolution. He was never able to explain it.

In his 1859 book **On the Origin of Species**, Darwin discussed the inexplicable lack of earlier fossils as one of the main difficulties for his theory of descent with slow modification through natural selection.

*So, given this history of life on Earth, it could be that **whatever caused the development of complex life from simple life forms, is a rare event and may be unlikely to occur on other planets. This finding, which violates our intuitions, does not bode well for discovering extraterrestrial intelligence.***

On the other hand, the law of large numbers could still play a role in finding other intelligent life out there. In spite of the pessimistic view described above, there still could be a very small fraction of exoplanets, out of the 100 billion, that had the right conditions fall into place at the right times and developed intelligent life in spite of the ultra-low probabilities.

So, there is still a small hope that we are not alone. There is much effort going into trying to find traces of their civilizations. *It is called the search for extraterrestrial intelligence, or SETI.*

SETI: The Search for Extraterrestrial Intelligence

People have long been interested in the idea that there are civilizations on other worlds. It had its start with Copernicus and his demonstration that the Sun did not revolve around us. Instead, he found that Earth was a planet revolving around the Sun. And that there were other neighboring worlds, like Mars and Venus.

The idea of different intelligences being out there came dramatically into the national consciousness in 1938. It was late October and the Mercury Theatre on the Air, a CBS radio program, dramatized H.G. Wells novel, **The War of the Worlds**.

Orson Welles, the twenty-three-year-old creator of the show, decided to make it very realistic by imagining it as a live radio news broadcast. Millions of people around the country, when they tuned in, thought they were listening to an actual news event and panicked into believing that the United States was actually being invaded by beings from Mars. It became the most famous radio show of all time.

Orson Welles

Many of the themes of science fiction books, movies, etc. have involved extraterrestrial intelligence. Carl Sagan was a popular astronomer who wrote the novel, "Contact." It was about the Earth's first interaction with a remote, alien civilization. The book was made into a movie in 1997 that starred Jodie Foster as the SETI astronomer who makes the contact.[CXI] *The popularity of the movie made many more people aware of the possibility of intelligent life beyond Earth.*

The *scientific interest* in the search for extraterrestrial intelligence, however, had its beginning in 1959 when physicists Giuseppe Cocconi and Philip Morrison of Cornell University published an article in the journal, **Nature,** with the provocative title, "Searching for Interstellar Communications." [255]

Cocconi and Morrison argued that radio telescopes had become sensitive enough to pick up transmissions that might be broadcast into space by civilizations on exoplanets, that is, planets outside of our solar system.

Such messages, they suggested, *might be transmitted at a wavelength of 21 cm (1,420.4 MHz).* This is the wavelength of radio emission by neutral hydrogen, the most common element in the universe. *They reasoned that those other intelligences might see this wavelength as a logical landmark in the radio spectrum and transmit over it in order to make contact.*

About the same time, Harvard astronomer, Harlow Shapley, tried to estimate *the number of inhabited planets in the universe.* It was a very rough, speculative calculation based on extremely limited knowledge at the time. He wrote:

> *"The universe has 10 million, million, million suns (10 followed by 18 zeros) similar to our own. One in a million has planets around it. Only one in a million million has the right combination of chemicals, temperature, water, days, and nights to support planetary life as we know it.*

> **"This calculation arrives at the estimated figure of 100 million worlds (in the Universe) where life has been forged by evolution."**

The publication of this calculation was another contributor to the idea among scientists that there should be many Earth-like planets out there with intelligent beings. But, again, the

---

[CXI] It is believed that Sagan based his main character on Dr. Jill Tarter, the chief scientist at the SETI Institute.

estimates and assumptions made by Professor Shapely were wildly speculative.

In 1961, just two years after the Cocconi-Morrison paper, Frank Drake, another Cornell professor, and an astronomer at the National Radio Astronomy Observatory[CXII] in Greenbank, West Virginia, was inspired by the paper, and began to use his radio telescope to search for intelligent signals. He called it "Project Ozma", after the Queen of Oz in L. Frank Baum's fantasy books.

Drake used the 85 foot diameter, Greenbank radio telescope to examine the stars Tau Ceti and Epsilon Eridani. He searched the segment of the spectrum called the "water hole" due to its proximity to the *hydrogen* and *hydroxyl radical* spectral lines, as suggested by Cocconi and Morrison.

A 400 kilohertz band around the marker frequency was scanned, using a single-channel receiver with a bandwidth of 100 hertz. He found nothing of interest.

Also intrigued with the discussions about the search for intelligent life, J.P.T. Pearman of the Space Science Board, a branch of the National Academy of Sciences, approached Dr. Drake and asked him to help convene a small, informal, SETI conference at the Green Bank observatory. The core purpose of the meeting, Pearman explained, was *to quantify whether SETI had any reasonable chance of successfully detecting civilizations around other stars.*

Drake accepted the idea. There was a total of ten scientists who attended the conference on November 1, 1961. Besides Pearman and Drake, Phillip Morrison, one of the authors of the paper that kicked off the idea, was also there.

Another attendee was John Lilly. He had conceived of the idea of learning *to communicate with dolphins* as practice for coming into contact with an alien intelligence.

[CXII] **The National Radio Astronomy Observatory** is a facility of the *National Science Foundation* operated under cooperative agreement by Associated Universities, Inc. Founded in 1956, the NRAO provides state-of-the-art radio telescope facilities for use by the international scientific community.

Perhaps the most famous attendee at the conference was another Cornell astronomer, Carl Sagan. He went on a few years later to write the best-selling book, **Intelligent Life in the Universe** (1966).

Carl Sagan

Bernard (Barney) Oliver, Director of R&D at Hewlett Packard, also attended. He had been involved in many of HP's electronic inventions. Oliver stayed with the SETI effort for the rest of his career.

Barney Oliver

Coincidently, I met Dr. Oliver when I was with the international consulting firm of Arthur D. Little, Inc. We were working for the National Science Foundation on a project to help them direct their research funds into the most productive areas of science and technology. Dr. Oliver shared his insights to me on this topic.

The famous "intelligent life equation" that bears Drake's name arose out of his preparations for this meeting.[256] He said:

"As I planned the meeting, I realized a few day[s] ahead of time we needed an agenda. And so, I wrote down all the things you needed to know to predict how hard it's going to be to detect extraterrestrial life. And looking at them it became pretty evident that if you multiplied all these together, you got a number, N, which is the number of detectable civilizations in our galaxy. This was aimed at the radio search, and not to search for primordial or primitive life forms."

The equation has *seven variables* that estimate *the number of active, communicative, extraterrestrial civilizations in our Milky Way galaxy.*

Professor Frank Drake

He knew that several of the factors in the equation were complete unknowns. But his purpose was to use it as framework for discussion during the conference.

This is the Drake equation:[257]

$$N = R_f \cdot f_p \cdot n_e \cdot f_l \cdot f_i \cdot f_c \cdot L$$

where:

N = the number of active, technological civilizations in our Milky Way galaxy with which communication might be possible.

That is, the planets with communicating civilizations which are on our current *past light cone*[CXIII];

and where:

R_f = the average rate of star formation in our galaxy,

f_p = the fraction of those stars that have planets,

n_e = the average number of planets that can potentially support life per star that has planets,

f_l = the fraction of planets that could support life that actually develop life at some point,

f_i = the fraction of planets with life that actually go on to develop intelligent life (civilizations),

f_c = the fraction of civilizations that develop a technology that releases detectable signs of their existence into space,

L = the length of time for which such civilizations release detectable signals into space.

Drake and his colleagues in 1961 made "educated guesses" for the *range of factor values:*

R_f = 1   There will be one star per year, on average, over the life of the galaxy; this was regarded as conservative by the group.

f_p = 0.2 to 0.5   This means that one fifth to one half of all stars formed will have planets.

n_e = 1 to 5   That is, stars with planets will have between 1 and 5 planets capable of developing life.

f_l = 1   This means that 100% of those planets will develop life.

[CXIII] **Past light cone** - *A cone-shaped portion of spacetime* containing all *past* locations from which light could arrive at a particular location within spacetime (the tip of the cone). Since nothing can travel faster than light, points outside the cone cannot in any way affect the tip.

$f_i = 1$ For each of these planets, 100% will develop intelligent life.

$f_c = 0.1$ to 0.2 10–20% of the planets with intelligent life will be able to communicate.

$L = 1000$ to $100,000,000$ -   these communicative civilizations will have a lifetime somewhere between 1000 and 100,000,000 years.

Inserting the set of "minimum" values for the estimate range for each factor into the equation produces a *minimum N of 20*. Incorporating the "maximum" values gives a *maximum of N = 50 million*. So, the range of N is 20 to 50 million.

The median of this range, assuming a normal distribution, is 25 million, active, communicating civilizations in our galaxy. **But again, most of the factors are complete unknowns, so this is a very shaky estimate.**

Although the attendees understood the limitations of their calculations, they convinced themselves that there was a good chance that the galaxy was teeming with life. *That encouraged them to go ahead and start looking more intensely for signs of the existence of those alien civilizations.*

Perhaps, because they were human, the participants had a bias toward wanting the answer to be strongly positive.

The Search for Exoplanets

Until the 1990s, we had no scientific evidence that planets outside our own solar system, called exoplanets, even existed. On 9 January 1992, radio astronomers Aleksander Wolszczan and Dale Frail announced the discovery of two planets orbiting the *pulsar* PSR 1257+12 (*not a main-sequence star*). This discovery was confirmed and is generally considered to be *the first definitive detection of an exoplanet.*[258]

The first confirmation of an exoplanet orbiting **a main-sequence star**[CXIV] was made in 1995, when a giant planet was found by Michel Mayor and Didier Queloz in a four-day orbit around the nearby star 51 Pegasi.[259] It is a hot "Jupiter" with a 4.2-day orbit. The discovery was made and confirmed with *the radial velocity method.*

This method for detecting exoplanets relies on the fact that a star does not remain completely stationary when it is orbited by a planet. Instead, the star wobbles, ever so slightly, in a small circle or ellipse, responding to the gravitational tug of its smaller companion. *Detecting this motion using an Earth based optical telescope is taken as confirmation of the planet's existence.*

A later, very productive method for identifying exoplanets was *NASA's Kepler space telescope mission. Kepler was a space craft designed to study thousands of stars in our Milky Way galaxy for signs of planets orbiting those stars.* It was launched in March 2009. The NASA scientists were especially interested in finding planets that might be suitable for Earth-like life.

The spacecraft found the planets, and some information about them, with a much different approach than the radial velocity method. *Kepler, instead, recorded the variability of the light coming from the star.*

If there was a planet orbiting the star, when the planet passed between the star and the Kepler spacecraft, it diminished the star's light being recorded on Kepler's extremely sensitive sensors.

The spacecraft then looked for the return of the same signal at a later time, corresponding to a full orbit of the potential planet around the star. In most cases for the exoplanets found, the return signal was measured in days, with the number depending on the size of the particular planet's orbit. They have ranged from 2.2 days to 242 days. In comparison, Earth orbits the sun every 365 days.

[CXIV] **A main sequence star** is any star that is fusing hydrogen in its core and has a stable balance of outward pressure from core nuclear fusion and gravitational forces pushing inward.

Kepler ran out of fuel and was retired in 2018 after nine years of service. Kepler's successor, the *Transiting Exoplanet Survey Satellite,* (TESS), was launched in 2018 and is now scanning a part of the sky which has at least 200,000 nearby stars. It is using the same *transiting technique* as employed by Kepler to find exoplanets.

According to NASA, as of October 16, 2021, there are *4,531 confirmed exoplanets* in 3,363 planetary systems, with 783 systems having more than one planet.[260]

However, there are 7,798 "candidate" exoplanet detections that still require further observations in order to confirm them. There are 97 exoplanets listed as confirmed by NASA within *10 parsecs[CXV]* or about 33 light years from Earth.

The vast majority of the exoplanets were found with *the transit methodology of Kepler and its successor, TESS.* About 1400 were found with the radial velocity method or other techniques.

Once a new planet is found and confirmed, other scientists study it to learn more about its characteristics, including whether our type of life might be able to exist on it.

For example, in 2008, scientists analyzed the light coming from the exoplanet HD 189733 to determine the chemical components found in the planet's atmosphere. *It was the first time an organic compound, methane, was found on an exoplanet, along with water vapor and carbon dioxide.*

These discoveries are a sign to the scientists that life, as we know it, could exist on an exoplanet that has conditions similar to those found on Earth. The probability of finding intelligent life, therefore, has increased.

About 4% of the exoplanets were found to be Earth-like in their characteristics. To be similar to Earth, the planet has to be a solid, rocky mass and have an orbit around the star that is at the right

---

[CXV] **Parsec** - a *unit of distance* used in astronomy, *equal to about 3.26 light years.* One parsec corresponds to the distance at which the mean radius of the earth's orbit subtends an angle of one second of arc.

distance to have surface temperatures where water can exist in its liquid form. *This is the so-called Goldilocks' zone.*

The Mystery of Tabby's Star

One of Kepler's discoveries has created some excitement about the potential for finding proof of extraterrestrial intelligence. One particular star, designated KIC 8462852 (which is about 1,500 light-years from Earth), *exhibited extremely strange results during Kepler's measurements.*

It is also nicknamed "Tabby's star," for Louisiana State University astronomer, Tabetha Boyajian, who led the team that first detected the star's unusual fluctuations.

Most of the planets discovered by the Kepler space craft showed less than one percent diminishment of the star's light as the planet passed by. Some of the signals seen over the last few years from Tabby's star, however, *diminished the starlight up to 22 percent, an enormous decrease compared to the 1 percent diminishment of even an exceptionally large, Jupiter-size planet.* The signals were also erratic in terms of their period.

Scientists considered many natural phenomena as explanations for the odd, unexpected results, like comet debris in orbit around the star, but none seemed to explain the observations. There are still some natural phenomena explanations to be explored. *But now the astronomers are seriously considering the remote possibility that the cause is a megastructure created by an intelligent, alien civilization.*

Such megastructures, possibly a super-sized solar array, have been theorized to be the method by which advanced civilizations would obtain power. They would capture large amounts of the energy released by their star using "mega" versions of solar power satellites.

This idea was popularized by Freeman Dyson, a professor at the Institute for Advanced Study, Princeton University. In his 1960 paper, "Search for Artificial Stellar Sources of Infrared Radiation"[261], Dyson speculated that such structures would be the logical consequence of the escalating energy needs of a technological civilization, and would be a necessity for its long-term survival. He proposed, therefore, that searching for such structures could lead to the detection of advanced, intelligent, extraterrestrial life.

If there is a mega-structure created by an alien intelligence around Tabby's star, then the aliens may also be sending out radio, or other signals, that can also be detected. One of the SETI groups currently engaged in examining the star for such signals is the SETI Institute of Mountain View, Calif., a private non-profit organization. They are using their Allen (Radio) Telescope Array (ATA) to search for non-natural radio signals coming from the vicinity of Tabby's star.

In case the potential civilization around KIC 8462852 is not broadcasting on radio frequencies, the SETI Institute is coordinating with the relatively new, Boquete *Optical* SETI Observatory in Panama. This facility is searching for brief, but powerful, laser pulses from the area around the star that constitute an intelligent transmission from an alien civilization.

So far, no signals have been detected.

There are other possible explanations for Tabby's star's behavior that are being explored. One is that there is a fine, dust cloud that is in orbit and periodically passing through and absorbing the light between the star and Kepler. *There is some evidence for such a phenomenon.*

Astronomers from the University of Arizona studied Tabby's star and *noticed that the dimming was more pronounced in ultraviolet (UV) than infrared light.* Any object bigger than a dust grain would cause uniform dimming across all wavelengths, study team members said.

> *"This pretty much rules out the alien megastructure theory, as that could not explain the wavelength-dependent dimming,"* lead author Huan Meng of the University of Arizona said in a statement. *"We suspect, instead, there is a cloud of dust orbiting the star with a roughly 700-day orbital period."*[262]

Eventually, we will learn the true cause of the excess diminishment of KIC 8462852 light. *All of the scientists agree that the chance is remote that the erratic results of the Kepler space craft observations will be explained by the presence of alien-built, megastructures around the star*. But, until they come up with an explanation of a natural phenomenon that fits the data, and that explanation is confirmed, some scientists will continue to retain the alien megastructure concept as a working hypothesis.

The Fermi Paradox

Enrico Fermi was an Italian American physicist who was a pioneer in nuclear physics. He headed the team that designed and built the first power-producing, nuclear reactor in 1942. He was also a key member of the Manhattan Project that created the atomic bomb.

After World War II, Fermi continued to work at Los Alamos National Laboratory in New Mexico, home of the Manhattan project. In 1950, he was having lunch with some colleagues, including Edward Teller, the "father of the hydrogen bomb". The group was discussing a New Yorker cartoon showing aliens emerging from a spaceship.

Fermi had done some rough calculations on the probability of intelligent life in the universe. He concluded that such beings were indeed highly probable. *"But", he said, "where are they?"*

Dr. Enrico Fermi

What Fermi meant is that intelligent aliens, if they existed, have had billions of years to colonize the universe and should have left visible signs of their presence everywhere. But all observations of the cosmos, even up to the present day, show a "dead" universe.

This has come to be known as the Fermi Paradox and has been a key point of discussion in astrobiology ever since.

Now, more than seventy years after Fermi made his statement, we have learned a great deal more about the possibilities. Thanks to our observations from space-based instruments, like NASA's Kepler spacecraft, we know with certainty that, in our Milky Way galaxy alone, there are estimated to be over 100 billion planets. About 1% of them, or 1 billion, would be Earth-like in size and composition.

*Over the **fourteen billion years since the big bang**, there should have been many civilizations that formed millions, or even billions, of years before us.* Given our "rapid" progress with space travel capability, it is reasonable to imagine that these beings had plenty of time to have created the means to easily travel to other stars.

After all, it only took us 65 years to go from the Wright brothers at Kittyhawk to Neil Armstrong stepping on the moon. And now, just fifty-three years after the moon landing, we[CXVI] are planning on landing men and women on Mars and setting up a major colony, consisting of thousands of individuals, within the next decade.

If we look back at human history and recognize that expansion and colonization are prime directives for our species. It is highly probable that if an advanced civilization like ours developed close to light-speed, space travel capability, their civilization would have exploded into the galaxy. Some calculations show that it would only take a brief, few million years to fill the galaxy with our presence.[263]

Yet, again, there is no sign of the civilizations that came before us.

Over the years, many scientists have proposed explanations for Fermi's Paradox. Professor John Ball of the Harvard-Smithsonian Center for Astrophysics suggested a possible answer in 1973.[264]

The aliens are out there, he said in his Icarus article, and they know about us, and other civilizations like us. But they have chosen to keep us in a type of preserve and let us mature and

---

[CXVI] NASA ,in conjunction with Elon Musk's SpaceX, are developing the detailed plans and means to carry them out.

evolve naturally. They will watch until they think we are ready to join the extraterrestrial community. And then they will contact us.

Another scientist with an explanation for the Fermi Paradox was Carl Sagan. In 1985, he commented on the lack of results from the Search for Extraterrestrial Intelligence (SETI) projects.

Since 1961, the date of the first SETI experiment, most of these projects have been using radio telescopes to try to detect signals from an alien civilization. And they have yet to hear the first whisper. *Sagan said that we may not be listening for the Aliens in the way they are broadcasting.* Their technologies are so advanced, that their communication systems are unknown to us. *Therefore, we should not expect to find them in the radio band.*

The Great Filter Theory

One of the key academic centers for considering the answer to the Fermi Paradox is the *Future of Humanity Institute at Oxford University in England.* Professor Robin Hansen, an associate professor of economics at George Mason University in the United States, and a research associate at the Future of Humanity Institute, proposed the "Great Filter" theory in 1998 to explain the lack of evidence for alien civilizations.[265] Many other scientists have picked up on his ideas and expanded them.

Robin Hansen

Essentially, Hansen said that *there are several steps of development* that aliens on another planet would have had to go through *to reach the point where their civilization could colonize the universe.* They include:

1. Their Earth-like planet must be near the "right" star system, like our sun, in the Goldilocks zone, and in the Galactic habitable zone,

2. Reproductive material (for example, RNA) must develop, then

3. Simple (prokaryotic) single-cell life develops, then

4. Complex (archaeatic & eukaryotic) single-cell life, then

5. Sexual reproduction, then

6. Multi-cell life, then

7. Tool-using animals with big brains, then

8. A high technology civilization develops to the level we are now, and then

9. Colonization explosion takes place.

So far, nothing among the trillions of stars in our universe has made it all the way along this path. *The "Great Silence" that we see in the visible universe implies that one or more of these steps are very improbable.* Therefore, *there must be a "Great Filter" along the path between simple, dead stuff and explosive life in the universe.*

This Great Filter stopped thousands of possible civilizations in our galaxy. So, they were never able to colonize the stars and that is why we do not see evidence of them.

The big question: Is that Great Filter behind us here on Earth, or ahead of us?

If it is behind us, that is good news. Earth has won the lottery. Everything necessary for intelligent life to develop and evolve came into place, even though all the probabilities were against it. Now, we will be the first civilization to develop interstellar space travel and, over the next billion years, spread throughout the cosmos.

On the other hand, *if the Great Filter is ahead of us*, then there were probably thousands or even millions of life forms before us, all of which hit the Great Filter and were destroyed before they could reach other stars.

How will we know which alternative is correct? If the planned NASA and SpaceX mission to Mars in the 2030s finds nothing, then the question is still unanswered.

If, on the other hand, the mission discovers the remnants of an ancient civilization, that would be bad news indeed. It would confirm that *the Great Filter is ahead of us, and as Robin Hansen says, "We're next!"*.

The Results of the SETI Program

NASA joined in SETI efforts at a low-level in the late 1960s. In 1971, NASA funded a SETI study that involved Frank Drake, Barney Oliver of Hewlett-Packard Laboratories, and others. The resulting report proposed the construction of an Earth-based radio telescope array with 1,500 dishes known as "Project Cyclops". The price tag for the Cyclops array was 10 billion dollars. *Cyclops was not built, but the report formed the basis of much of the SETI work that followed.*

On August 15, 1977 a strong, narrow band signal, of the type SETI astronomers had been seeking, was recorded by The Ohio State University's Big Ear, radio telescope. The signal appeared to come from the direction of the constellation Sagittarius and bore the expected hallmarks of extraterrestrial intelligent origin. It was an extraordinarily strong signal but was not modulated. It lasted for 72 seconds.

Astronomer Jerry R. Ehman discovered the anomaly a few days later while reviewing the recorded data. He was so impressed by the result that he circled the reading on the computer printout and wrote the comment *"Wow!"* on its side. This became the name attached to the event.[266]

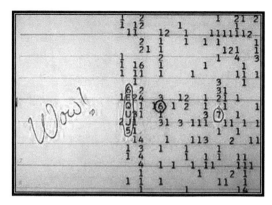

The "Wow!" signal

Despite numerous attempts, *no one has ever been able to receive that signal again.* The Wow! Signal appears to have been a one-time event and is, therefore, not considered to be proof of an alien civilization. *It is still, however, the strongest candidate for an alien radio transmission ever detected*

A major step-up in SETI technology took place in the Boston area. In 1981, Harvard University physicist, Paul Horowitz, designed and built a spectrum analyzer specifically intended to facilitate the search for SETI transmissions. It had a capacity of 131,000 narrow band channels and expanded dramatically the number of channels that could be monitored. Later, he increased the analysis hardware to one billion channels.

The technology was used with the 85 foot diameter, Harvard/Smithsonian radio telescope at Oak Ridge Observatory in Harvard, Massachusetts. This project was named "Sentinel".

It was about this time that Stephen Spielberg, the famous Hollywood director of films like "ET" and "Close Encounters of the Third Kind", along with Carl Sagan's Planetary Society, made major donations to the Sentinel project. They viewed it as having a good chance of making contact.

At this time, I was living in the town of Harvard, Massachusetts where the Harvard Observatory is located. I had an opportunity to take an Astronomy course there. As part of the program, we were shown the banks of computers and racks of analysis hardware. The tour included a briefing on the SETI technology and the process for their search.

In addition to the SETI project, the Oak Ridge observatory was among the first to discover an exoplanet using a large optical telescope and the radial velocity or "wobble" methodology.

In later years, Professor Horowitz continued to improve the SETI technology. Unfortunately, a strong windstorm in August 2005 destroyed the radio antenna and the observatory was closed.

Dr. Paul Horowitz

On Columbus Day in 1992, NASA started a formal, more intensive, SETI program. Less than a year later, however, Congress became disillusioned with the effort and canceled the program. This was the end of government support for the endeavor.

Today, there are still many active, SETI operations around the world, searching for signs of intelligent civilizations beyond our own. Some of these organizations are supported by universities, others by governments and still others by private donations. Two examples of the academic type, other than Harvard University, are the programs at Princeton University and the University of California at Berkeley.

The SETI Institute, a non-profit organization in California, has become the spearhead for SETI in the United States. They were formed when NASA cancelled their contract in 1993.

The Institute relies primarily on private funding. It started with just two employees, founder Tom Pierson, a former grants administrator at San Francisco State University, and astronomer, Jill Tarter. She is still active in the SETI programs and is a key spokesperson for their efforts.

Over the years, other research disciplines have been added to the Institute's portfolio, all unified by their relevance to the search for, and understanding of, life beyond Earth. Today, the Institute has approximately one hundred scientists as well as specialists in administration, education, and outreach.

The most ambitious SETI technology development by the institute is the Allen Telescope Array (ATA) located in Shasta County, California, northeast of San Francisco. It became operational in 2007 with 42 of its 350 planned radio antennas in place. Unfortunately, it has been beset by financial problems and has only operated on and off over the ensuing years.

The ATA runs independently. If a signal that it receives appears to be of intelligent origin, it is identified by the *automated, analysis system*. The system immediately alerts the scientists to check the characteristics of the reception and confirm its origin.

In spite of significant work and highly improved technology to conduct the searches, more than sixty years of SETI effort has not discovered any confirmed sign of intelligent life elsewhere in the galaxy. The SETI scientists, however, are not discouraged and continue their work.

The Drake Equation Revisited

Recently, Anders Sandberg, a colleague of Robin Hansen and Nick Bostrom at Oxford's *Future of Humanity Institute*, took another look at Drake's equation and updated it with more recent knowledge of biology, chemistry, and cosmology. And then dealt with the remaining uncertainties in our knowledge using sophisticated, statistical methods.[267]

Anders Sandberg

His conclusion was enlightening. Fermi's paradox arises when we combine a *high and extremely confident estimate of the number of civilizations in our galaxy with the absence of any evidence for their existence.* But the high confidence that causes this clash results from *applying the Drake model and* **using point estimates that are more certain than warranted** *for the parameters.* Sandberg said:

> *"These estimates, however, make implicit knowledge claims about processes (especially those connected with the origin of life) which are untenable given the current state of scientific knowledge.*
>
> *"When* **we take account of realistic uncertainty, replacing point estimates by probability distributions** *that reflect current scientific understanding, we find no reason to be highly confident that the galaxy (or observable universe) contains other civilizations, and thus no longer find our observations in conflict with our prior probabilities (i.e. the Fermi Paradox).*
>
> *"We found qualitatively comparable results through two different methods: using the authors' assessments of current scientific knowledge bearing on key parameters and using the divergent estimates of these parameters in the astrobiology literature as a proxy for current scientific uncertainty.*
>
> *"When* **we update this prior in light of the Fermi observation, we find a substantial probability that**

we are alone in our galaxy, and perhaps even in our observable universe (53%–99.6% and 39%–85% respectively).

"'Where are they?' — probably extremely far away, and quite possibly beyond the cosmological horizon and forever unreachable."

Sandberg, et al's calculations, therefore, had a surprising conclusion. **The chance that humanity is alone in the Milky Way galaxy, has a median probability of about 76%.** *In other words, there is a particularly good chance that E.T. does not even exist.*

So, they say, we should not be surprised that we do not see evidence of Aliens. And, therefore, *there is no paradox. We are just the lucky ones.*

But, most scientists agree, *we still do not have enough information to make an absolute conclusion.* Given Sandberg's calculations, there is also a 24% chance that E.T. *is* out there.[CXVII] So, the SETI scientists and the astrobiologists should continue with their work.

[CXVII] The total probability of an event occurring or not occurring is 100%, either it does, or it does not. If the probability is 76% that there is no active civilization in the galaxy, then there is a 24% chance that there is one.

UFOs and Extraterrestrials

There is a long history of people seeing Flying Saucers and some even reporting that they have interacted with the occupants. *To this day, people are wondering if intelligent extraterrestrials are visiting us.*

This idea came into being during the late 1940s after a series of well publicized incidents. *The military and some government agencies agreed that something was going on.* Then, over the next ten to fifteen years, the Pentagon's attitude changed.

The government policy of the 1960s and beyond became one of denial and derision of the existence of UFOs. That attitude lasted for almost fifty years after the 1969 close of the Air Force's Project Blue Book, the last official government UFO investigation.

Then in 2017, the tide turned once again. The venerable New York Times published an article that described fighter planes based on the aircraft carrier, USS Nimitz[268] encountering enigmatic UFOs off the coast of California. The confrontations took place in 2004.

Unauthorized videos were released showing the encounters on the fighters' weapons systems displays. The Pentagon later confirmed that the videos were authentic.

The U.S. government's acceptance of UFOs as real objects officially occurred on June 25, 2021, with the release of a congressionally mandated, Pentagon report describing 140 military cases, *all but one unexplained.* The military encounters from 2004 to 2021 that were discussed in the report were similar to those that occurred with the Nimitz based aircraft.[269]

Unknown to the public, after the Nimitz encounter in 2004, the Pentagon had reinstituted a UFO study group in the Pentagon. It was called *the Advanced Aerospace Threat Identification Program* (AATIP). The results of their investigations formed the basis of the 2021 Pentagon Report.

The bottom line of the report: for the first time in history, the Pentagon brass says that UFOs, or Unidentified Aerial Phenomena (UAPs) as they now call them, are very real and very mysterious. **They cannot say if the UAPs are Alien in origin, but they cannot rule it out either.**

How the Phenomena of Flying Saucers and Extraterrestrials Began

I grew up in the 1950s when "Flying Saucers" were often in the news. There were numerous B-movies produced featuring the saucers and the creatures that went with them. This helped to further fuel the attention and concern about these mysterious entities.

The idea of "Flying Saucers" had their beginning on June 24, 1947. A private pilot named Kenneth Arnold was on a business trip flying his airplane near Mount Rainer, in the state of Washington. Suddenly, he noticed a string of nine, shiny objects flying at speeds that Arnold estimated to be about 1,200 miles an hour.

Having had a long career in aviation, Arnold was a credible witness. He described the objects as having convex shapes. In his many interviews with the press following his sighting, he compared the objects' shapes to a "disc", "half-moon", or generally "convex and thin".[270] The photo below shows his conception of the object's shape, and it is clear that it is not a "saucer".

Kenneth Arnold

It was the press that started calling the objects "Flying Saucers" while reporting on the wave of many other sightings that followed the publicity of the Arnold report.

The saucer idea probably came from Arnold describing the *movement* of the objects as *"saucers skipping on water" or* having a kind of wobbling motion.

Over the next month, several more sightings were made in the Mount Rainer area. One sighting of particular interest occurred a couple of weeks after Arnold, on July 4, 1947. A United Airlines crew, on route to Seattle, also spotted five to nine disk-like objects that paced their airplane for ten to fifteen minutes before suddenly disappearing. The report of this sighting received even more newspaper coverage than Arnold's original observation and opened the floodgates of media interest.

Over the next several weeks, the idea of "Flying Saucers" burst across the country and the world. There were several hundred reports of similar sightings from various parts of the United States and other countries. Most of the reports described *saucer-shaped objects. The observers may have been influenced by the mistaken use of the term by the media when talking about the initial Arnold sighting.*

It was Kenneth Arnold in some of his later interviews that introduced the idea that the objects could be extraterrestrial in origin. From a July 7, 1947, article in the Chicago Times:

> *"....he said discs were making turns so abruptly in rounding peaks that it would have been impossible for human pilots inside to have survived the pressure.*[CXVIII] *So, he too thinks they are controlled from elsewhere, regardless of whether it's from Mars, Venus, or our own planet."*[271]

On July 8, 1947, another event occurred which had a similar impact on the country as Arnold's original sighting. The Roswell (New Mexico) Army Air Field (RAAF) public information officer, Walter Haut, issued a press release stating that personnel from the field's 509th Operations Group had recovered a "flying disc", which had crashed on a ranch near Roswell. The report was immediately picked up by numerous news outlets around the country.[272]

[CXVIII] **Note**: this is exactly the same observation as reported by the Navy Pilots in the "Tic-tac" sightings in 2004 through 2017. And by so many other UFO reports over the years.

It is interesting to note that the 509th Operations Group, based at Roswell at that time, traced its heritage back two years to when its B-29 Superfortress bombers dropped their atomic bombs on Japan. An action that helped end the war in the Pacific.

*This fact is of note since, over the years, and even to this day, many Flying Saucer sightings in the United States, as well as various other countries, continue to be **associated with areas around or related to our nuclear weapons facilities**.* [273]

Colonel William H. Blanchard, commanding officer of the 509th at Roswell, notified his superior, General Roger M. Ramey of the Eighth Air Force in Fort Worth, Texas, of the crashed saucer. Ramey ordered the recovered object to be flown to Fort Worth Army Airfield.

Ramey later released a statement in a press conference identifying the object as being a weather balloon and its "kite", a nickname for a radar reflector used to track the balloons from the ground.[274]

The Roswell incident was then pretty much forgotten. *But it became a cult favorite when it was revived thirty years later in a 1978 book by UFO researcher, Stanton Friedman.*[275]

One of Friedman's interviews for his book was Major Jesse Marcel, the Intelligence Officer at the RAAF at the time of the incident. Marcel was the only person known to have accompanied the Roswell debris from its recovery site on the ranch owned by Mack Brazel, to General Ramey in Fort Worth.

In a 1979 documentary movie, Marcel described his participation in the 1947 Fort Worth press conference:[276]

> *"They wanted some comments from me, but I wasn't at liberty to do that. So, all I could do is keep my mouth shut. And General Ramey is the one who discussed – told the newspapers, I mean the newsman, what it was, and to forget about it. It is nothing more than a weather observation balloon. Of course, we both knew differently."*

Major Jesse Marcel with Roswell "debris"

There is now a Flying Saucer Museum in Roswell, opened in 1992. It was started by some of the "witnesses," including Walter Haut, the Army Air Force officer who wrote the press release announcing the 1947 crash.

I visited the museum when I passed through Roswell and had an opportunity to interview one of the other witnesses, Glen Dennis. He was also a founder of the museum.

Dennis was employed by the local funeral home at the time of the incident. He claims that he was approached by an officer from RAAF who wanted to order several, "small", child-sized, hermetically sealed coffins.

Glenn Dennis

In his side job of ambulance driver for Roswell, Dennis brought an airman, who was injured in an auto accident, to the base hospital. After he dropped the airman off, he bumped into a nurse who was a good friend.

She excitedly told him that she had just been involved in the autopsy of several, small alien bodies taken from the crash site, and she was warned to forget it.

When Dennis tried to contact her a few days later, he found that she had "disappeared." No one seemed to know where she had gone. Dennis himself claims that he was threatened by a big, "red headed" Army Captain if Glenn disclosed what he knew.

With the later, high interest in the Roswell incident, Glenn Dennis' testimony has been very thoroughly investigated, and many researchers found it to be flawed. No one was ever able to find any trace of his friend the nurse who would have been a key witness.

Conspiracy theories about the Roswell event nevertheless persist, and the incident continues to be of high interest in popular media. It has been described as "the world's most famous, most exhaustively investigated, and most thoroughly debunked, Flying Saucer claim".[277]

However, the incident, along with all of the other Flying Saucer sightings that occurred in 1947, *prompted the U.S. to launch serious investigations that lasted for more than twenty years.*

The USAF Investigations, 1948-1963

On July 2, 1926, the Congress created the *Army Air Corps* as the flying wing of the Army. But the National Security Act of 1947 established an entirely separate branch, the *United States Air Force (USAF)*. So, *the Air Force officially came into being on September 18, 1947.*

Thus, it was the USAF that assumed the responsibility to investigate the Flying Saucers. On September 23, 1947, Lieutenant General Nathan Twining,[CXIX] then the Commander of Air Material Command (AMC), issued a memo to Brigadier General George Schulgen. The subject line of the memo read "AMC Opinion Concerning 'Flying Discs.'"

The overall tone of the memo was that the unidentified objects seen in the skies by military personnel *were not weather, astronomical, or other phenomenon* but rather actual objects that warranted further investigation. Twining wrote *"The phenomenon reported is something real and not visionary or fictitious."*

Lieutenant General Nathan Twining

Project Sign (1948-1949) (originally named Project Saucer) was officially launched in early 1948. It was the first official, military-

---

[CXIX] Twining later became the Air Force Chief of Staff in 1953, upon the retirement of General Hoyt Vandenberg.

intelligence program to collect information on Flying Saucer sightings. Project Sign was organized under the Air Force Director of Intelligence, General Charles Cabell. The Project Director of Sign was Captain Robert R. Sneider.

After looking at numerous reports and conducting detailed intelligence investigations over the next year, Sneider, and his team *favored the hypothesis that the Saucers were extraterrestrial, as the best explanation for their origin.* This was obviously a daring conclusion to reach and Sneider and his team must have felt strongly about their evidence.

As part of the required reporting for *any* Air Force Intelligence investigation, the Sign team had to write an *"Estimate of the Situation"* as part of their final report on Project Sign. It was the equivalent of an "Abstract" used for journal articles. The goal of the Estimate was to present a "best judgement" summary of the conclusion of the study, strongly backed with as much "proof" as possible.

The Estimate and the full report were sent to Director of Intelligence, Lieutenant General Charles Cabell. Cabell apparently, had no problem with the findings, but saw the issue as a "hot potato". So, in turn, he sent the report on to Air Force, Chief of Staff, General Hoyt Vandenberg.[278]

Lieutenant General Charles Cabell

Vandenburg was quick and decisive. With a noticeably short time for contemplation, he sent the Estimate back down the chain to Cabell, and then Sneider. He termed the Estimate's conclusions as "unacceptable". He did not want an extraterrestrial assessment for Flying Saucers. *He was saying clearly and loudly to Sign, and everyone in between, that he was not happy with this as an answer.*

General Hoyt Vandenberg

*Project Sign was then dissolved and was replaced by **Project Grudge** (1949-1951) with the same mission, to investigate the Flying Saucers.* At first, there was just a change in code names, but eventually Sneider and his team were replaced.

Many critics charged that, *from its formation, Project Grudge was operated under a directive to debunk the sightings.* It is not clear where exactly this directive originated but it did not come from General Cabell, the head of the Intelligence command.

All Saucer reports were judged by the Grudge team to have prosaic explanations, even though little research was ever conducted. Some of the "evaluations" produced under Grudge were awkward or even logically unsustainable.

Dr. Michael D. Swords wrote in a book about Grudge: [279]

> *"Inside the military, Major Aaron J. Boggs in the Pentagon and Colonel Harold Watson, at Air Material Command, were openly giving the impression that the whole Flying Saucer business was ridiculous. Project Grudge became an exercise of derision and sloppy filing."*

Not all high-level officers in the Intelligence command, however, wanted to downplay the Saucer reports. Some were indeed still concerned about their origin and their intent.

When USAF Director of Intelligence, General Charles Cabell learned that Grudge had essentially ignored many Saucer reports, he became furious. At a meeting, a frustrated Cabell was reported to have said, [280]

> *"I want an open mind; in fact, I order an open mind! Anyone who doesn't keep an open mind can get out now! ... Why do I have to stir up the action? Anyone can see that we do not have a satisfactory answer to the Saucer question."*

At another meeting, this one with a high-ranking group of Colonels, Cabell said, *"I've been lied to, and lied to, and lied to. I want it to stop. I want the answer to the Saucers, and I want a good answer."*

Cabell characterized the 1949 Grudge report as "tripe".

In his 1956 book, **The Report on Unidentified Flying Objects**, Edward J. Ruppelt, Director of the later Saucer investigation, Project Blue Book, described Grudge as the "Dark Ages" of USAF, Saucer investigations. Grudge's personnel were in fact conducting little or no analysis, while simultaneously saying that all Saucer reports were being thoroughly reviewed.[281]

Ruppelt additionally reported that the word "Grudge" was chosen deliberately by the "anti-Saucer" elements in the Air Force. The meaning was considered to be that the Air Force had a "grudge" against Saucers. Everything was being evaluated on the premise that Saucers could not exist. *They were saying, "No matter what you see or hear, don't believe it."*

By the end of 1951, General Cabell and other Generals in his command were so dissatisfied with the state of the Air Force Saucer investigations that, *in March 1952, they dismantled Project Grudge and replaced it with **Project Blue Book (1952-1969)**.* Blue Book would turn out to be the longest lasting of the series of Saucer investigations by the military.

One of the other Generals on Cabell's staff that supported the decision was William Garland. He thought that the Saucer question deserved serious scrutiny because he had witnessed a Flying Saucer himself.[282]

The new name, Project Blue Book, was selected to refer to the blue booklets used for testing at many colleges and universities. The name was inspired, said Ruppelt, by the close attention that high-ranking officers were giving the new project. They wanted the study of Saucers to be as important as a college final exam.

Blue Book was headquartered, as were the earlier Projects, at Wright-Patterson Air Force Base near Dayton, Ohio. The director of the program was Captain Edward J. Ruppelt, a USAF officer who had also been associated with Grudge. He was, however, viewed by his superiors as being much more open minded about Flying Saucers than those officers initially in control of Grudge.

Captain Edward J. Ruppelt

The two goals given to Project Blue Book were to (1) *determine if Flying Saucers are a threat to national security, and (2) to scientifically analyze the Saucer related data.*

Up until Project Blue Book, the official term for the objects was Flying Saucers. Captain Ruppelt believed they needed a more general term to replace "Flying Saucers" because *there were so many different shapes being reported.*[CXX] He created "Unidentified Flying Object", or UFO, as the new term and it stuck for the next six decades.

[CXX] Common shapes considered to be exotic (not man made) include sphere, disc, oval, diamond, rectangle, triangle, egg, and teardrop.

Ruppelt implemented a number of other changes: He streamlined the manner in which UFOs were reported by military officials, partly with the hope of alleviating the stigma and ridicule associated with UFO witnesses.

By now, there was a deep, underlying, national interest in UFOs. **Life** magazine, a large circulation publication at the time, printed an article on UFOs in April of 1952 that added to the sensation. Its title was *"Have we visitors from space?"*[283]

The article, *written with Ruppelt's full cooperation*, explained the Air Force's national-security interest in UFOs. The piece also made a convincing case for the extraterrestrial origin of UFOs, through the colorful retelling of ten, unexplained UFO "incidents". As one rocket scientist who was working on "secret" projects for the U.S. told LIFE: *"I am completely convinced that they have an out-of-world basis."*

The Summer of 1952, the first year of Blue Book's operation, changed everything for the Country's, and the military's, view of UFOs. There were so many sightings over Washington, D.C. that it was known as "the Big Flap."

Major American newspapers were regularly reporting multiple, credible sightings by civilian and military radar operators and pilots.[284]

On July 19th, 1952, things took a turn for the worse. Radar operators at National Airport watched the objects buzz past the White House and Capitol building. Now the UFO situation became profoundly serious from a military point of view. The unknown objects with unknown intentions had approached the seat of the U.S. Government. *People inside and outside of the government were starting to panic.*

Two F-94 interceptor jets were scrambled by the Air Force on that day. But each time the fighters approached the locations of the UFOs as they appeared on the radar screens, the mysterious blips would disappear.

Finally, one of the jet pilots caught sight of a group of bright lights in the distance and gave chase.

> *"I tried to make contact with the bogies below 1,000 feet,"* the pilot later told reporters. *"I saw several bright lights. I*

was at maximum speed, but even then, I had no closing speed. I ceased chasing them because I saw no chance of overtaking them."

The newspaper headlines the next day screamed "Saucers Swarm Over Capital" and "Jets Chase D.C. Sky Ghosts." The public panic over the sightings was so great that President Harry Truman asked his aides to get some answers.

F-94 Interceptor

Before the Presidential Aides could develop their report, the Air Force called a high-level press conference. Because of all of the publicity and the extreme concern on all fronts, it was viewed as the most urgent, military presser since World War II.

The Air Force brass had decided, without consulting Ruppelt or the Project Blue Book team, that the best response to the sightings was to *feed the press and the public an easy-to-swallow explanation, whether or not it was true.*

The Generals had decided that they would say that the radar observations were caused by a weather phenomenon called "temperature inversion". The effect, they said, causes "blobs" to appear on the radar screens which explains the sightings. *This is the story that Major General John Samford gave in his famous press briefing of July 29, 1952.*

Major General John Samford at the UFO briefing

Samford spent most of the briefing time talking about the temperature inversion as being the cause of the D.C. sightings. However, he also said:

> *"It would be foolhardy to deny that higher forms of life exist elsewhere. It would be unreasonable to deny that we have already been visited by beings from Outer Space.*
>
> *"We believe that all of this will eventually be understood by the human mind, and it is our job to hasten this understanding."*

In commenting on the 20% of UFO sightings that have no explanations, he said:

> *"The Saucers behavior indicates that they either have unlimited power or no mass. Many credible people have seen incredible things."*

So, the Generals came up with an explanation for the particular group of sightings over the Nation's Capital and *did not dismiss the idea of the UFO phenomena.*

Captain Ruppelt was extremely disappointed. He and his investigators *had not been consulted before the briefing.* The Blue Book staff had, in fact, ruled out the inversion idea and yet it was being presented to the public as the major reason for all of the sightings. The radar operators themselves had disputed the possibility. They said:

> *"We know what inversions look like. This is not an inversion. This is not the same thing at all."*

Former radar controller, Howard Cocklin, told the Washington Post in 2002 that he was still convinced that he saw an object. *"I saw it on the [radar] screen and out the window" over Washington National Airport."*[285]

The papers, however, fell in line and reported the temperature-inversion story and the public largely seemed to accept it. UFO sightings dropped from fifty per day, down to ten. So, the Air Force brass had met their goal for the briefing.

The CIA, on the other hand, internally reacted strongly to the 1952 wave of UFO reports. As a result of the intense interest of President Truman and the Public, the Intelligence Agency formed a special study group within the *Office of Scientific Intelligence* (OSI) and *Office of Current Intelligence* (OCI) to review the situation.[286]

After this review of the data from Project Blue Book, the *CIA recommended* to the Intelligence Advisory Committee (IAC) in December of 1952 that *a scientific study of the UFO phenomena be undertaken.*

One of the major concerns the CIA cited was the possible overloading of U.S. Defense communication and response systems as the military became swapped with tracking the UFOs. **This concern became the basis of the major campaign to discourage public interest in Flying Saucers.**

In January 1953, the CIA instigated the formation of the *Robertson Panel*, to be headed by Howard P. Robertson, to deal with these issues. Secretly, the intent for this panel investigation was to begin the campaign to discredit saucer sightings.

Professor Robertson was a well-known physicist and mathematician from Princeton University. He was also a CIA consultant, and the director of the Defense Department's *Weapons Evaluation Group.*

Professor Howard P. Robertson

Robertson was instructed by the Office of Scientific Intelligence (OSI) to assemble a group of prominent scientists to review the Air Force's UFO files and report on their findings and conclusions. He brought together five other, top-level scientists to conduct the review. They included J. Allen Hynek an Ohio State University Astronomer and a major consultant to the Air Force's UFO studies.

The Robertson Panel's study was far from comprehensive. It only met on four consecutive days of formal meetings. In total, they were together for twelve hours. They reviewed just twenty-three cases out of the 2,331 cases on record. This amounted to only about 1% of the total. Edward Ruppelt, Director of Blue Book, later wrote in his hardcover, however, that the Panel did study Blue Book's best cases.[287]

The Panel concluded in their final Report:

1. The evidence presented on Unidentified Flying Objects shows *no indication that these phenomena are a direct, physical threat to national security.*

2. Most UFO reports could be *explained as misidentification of mundane aerial objects,* and the *remaining minority* could, in all likelihood, be *similarly explained with further study.*

3. The continued emphasis on the reporting of these phenomena, however, does pose an indirect hazard by *threatening to overwhelm standard, military*

communications due to the public interest in the subject and the over reporting of incidents.

4. A *public education campaign should be undertaken* in order to reduce the public's interest in the subject,

    a. This would minimize the risk of swamping Air Defense systems with reports at critical times,

5. The national security agencies need to take immediate steps to *strip the Unidentified Flying Objects of the special status they have been given and the aura of mystery they have unfortunately acquired.*

6. We suggest that these aims may be achieved by an integrated program designed to reassure the public:

    a. of the total lack of evidence of Inimical forces behind the phenomenon,

    b. to train personnel to recognize and reject false indications quickly and effectively[288],

The Robertson Panel's report was contained within a larger, internal CIA report by F. C. Durant, a CIA officer who served as Secretary to the Robertson Panel. *This wider document is commonly referred to as the Durant Report.*[289]

The Robertson recommendations helped shape Air Force policy for UFO studies, not only immediately afterward, but also until just recently. Reflecting the new attitude, the Air Force issued a *set of stringent controls on Blue Book's operations.* The staff of Blue Book could discuss UFO cases with the media *only if* the cases were regarded as having a conventional explanation.

If the cases were unidentified, the media was to be told only that *the situation was being analyzed.* Blue Book was also ordered to reduce the number of "unidentified" cases to a minimum.

Blue Book had essentially become a public relations outfit with a debunking mandate. For example, by the end of 1956, the number of cases listed as unsolved had dipped to barely 0.4 percent, from 20 to 30% only a few years earlier.

The key impact of the Robertson Report was that there was a significant decline in the concern and interest in UFOs in government agencies. J. Allen Hynek, a member of the Panel, lamented that the Report had:

"...made the subject of UFOs scientifically unrespectable, and for nearly twenty years, not enough attention was paid to the subject to acquire the kind of data needed even to decide the nature of the UFO phenomenon."

In his book, Ruppelt described the stripping of Blue Book's investigative duties following the Robertson Panel and the subsequent demoralization of his staff.

Eventually, Ruppelt requested reassignment. By the time of his departure in August 1953, his staff had been reduced from more than ten to just two subordinates and himself.

UFO researcher, Jerome Clark, wrote:

"Most observers of Blue Book agree that the Ruppelt years comprised the project's golden age, when investigations were most capably directed and conducted. Ruppelt was open-minded about UFOs, and his investigators were not known, as Grudge's were, for force-fitting explanations on cases."[290]

His temporary replacement as Blue Book Director was a noncommissioned officer. Most of the officers who later succeeded him as Blue Book Director showed either apathy or outright hostility to the subject of UFOs or were hampered by a lack of funding and official support.

Similarly, each general who oversaw Blue Book had a different administration, different goals, and different interpretations of what they were looking for, and sometimes, their findings disputed those of their predecessors.

So, Project Blue Book *descended into a new "Dark Ages"* from which many UFO investigators argue it never emerged.[291]

Following a 1965 wave of sporadic UFO reports, astronomer and Blue Book consultant, J. Allen Hynek, wrote a letter to the Air Force Scientific Advisory Board (AFSAB) *suggesting that a "civilian panel of physical and social scientists"* convene to re-examine if a UFO *"problem"* even existed[292].

The result was the formation of *The Condon Committee.* It was the informal name of the "University of Colorado UFO Project". This committee was funded by the United States Air Force from

Professor J. Alan Hynek

1966 to 1968 to study UFOs *under the direction of physicist, Edward Condon.*

Dr. Edward Condon

There was initially some concern among the administration of the University of Colorado about taking on such a project. Robert J. Low, an assistant dean of the University's graduate program, *and the Coordinator of the Condon Committee*, wrote a memo on August 9, 1966, to his administrators that *they could expect the study to show that UFO observations had no basis in reality.* A clear indication that this committee was also biased toward dismissing the saucer phenomenon.

The Low memo was discovered by some of the Committee members during the course of the study. They became concerned about the built-in bias in reporting the yet-to-be-determined results of the study.[293]

This concern about prejudice in the study was magnified with a speech given by Dr. Condon. In late January 1967, Condon said, in a lecture, that *he thought the government should not study UFOs* because *the subject was 'nonsense', adding, "but I'm not supposed to reach that conclusion for another year."*[294]

The result of the Condon Committee's work was formally titled, **Scientific Study of Unidentified Flying Objects**. The Committee delivered its Report to the Air Force in November 1968. It was 1,485 pages in its hardcover edition.[295] *The clear evidence of prejudice on the part of Condon, however, colored the results of the study for its critics.*

Condon wrote in his section on the Conclusions:

> *"Our general conclusion is that nothing has come from the study of UFOs in the past 21 years that has added to scientific knowledge. Careful consideration of the record, as it is available to us, leads us to conclude that further extensive study of UFOs probably cannot be justified in the expectation that science will be advanced thereby."*[296]

He also recommended *against* the creation of another government program to investigate UFO incidents.[297] *This report, therefore, was the informal end of Project Blue Book and government sponsored investigations of UFOs. It took another year, however, until the end of 1969, to make it official.*

The Formal Findings from Project Blue Book

From 1952 to 1969, the seventeen years that the Air Force investigations ran, they collected 12,618 UFO reports. 11,917 of them were explained away as the result of cloud coverage obscuring aircraft lights, classified Airforce training exercises, mirages in the deserts of the southwestern United States or a range of other natural phenomena. 701 cases, or about 6% of the total, remain "Unidentified."

In late 1969, Secretary of the Air Force, Robert C. Seamans, Jr. announced that Project Blue Book was ending, as there was no scientific evidence to prove that UFOs were a matter of national security. *The project officially ceased to exist on December 17, 1969,* though some research efforts continued for another month, until January of the following year.

The official findings of Project Blue Book claimed that four things influenced UFO sightings:

1.   Mass hysteria among the American people

2.   Individuals hoping to propose a hoax to seek fame.

3.   Psychopathological persons.

4.   Misidentification of conventional objects.

The major conclusions of the Project were:

1.   No UFO reported, investigated, and evaluated by the Air Force has ever given any indication of being a threat to our national security.

2.   There has been no evidence given to, or discovered by the Air Force, that sightings categorized as "unidentified" represent technological developments or principles beyond the range of present-day scientific knowledge.

3.   There has been no evidence showing the sightings categorized as "unidentified" are extraterrestrial vehicles.

In short, Project Blue Book claimed to have solved the UFO mystery "once and for all" by chalking it all up to natural phenomena.

UFO Sightings by Citizens

For almost seventy-five years, since Kenneth Arnold reported his encounter, ordinary people have also been reporting sightings of UFOs. The observers amount to about 6% of the population.

There are now about eight to ten thousand reports per year in the U.S. alone. The Mutual UFO Network (MUFON) is the largest private reporting agency that collects and tracks this data.

Most sightings, about 95%, which are investigated by MUFON can be explained as airplanes, balloons, the planet Venus, the space station, and other natural phenomena. *But about 5% are classified as UFOs.*

I have always enjoyed looking up at the sky to watch meteor showers, comets, eclipses, and other natural phenomena. Once, however, I saw something I could not explain. And it happened where I live in Gloucester, Massachusetts.

My wife and I were standing on our back porch on a warm August night in 2016, about 10 p.m. looking out over Ipswich Bay. Off in the distance, I saw two bright lights that I thought were landing lights on a commercial passenger plane. They were very bright. I had seen that phenomenon many times in the past, so I was not surprised.

But what caught my attention was that the lights were not white, the usual color, but orange. So, we continued to watch them.

The bright lights were heading toward us and eventually passed overhead, just in front of us. They were perfectly formed, bright, orange spheres or orbs, which looked slightly translucent with a brighter center. I estimated their altitude and speed to be that of a small, private, prop-driven airplane, or about 1000 feet, and 120 miles per hour.

What happened next was even more surprising. We saw more of the objects coming in a line in back of the first two, following the same course. I counted a total of twenty objects over the next twenty minutes as they went by, all looking and behaving exactly the same as the first two.

I had never heard of orange orbs being reported as UFOs. So, I did some online research. I found that it was a rather common description. There were hundreds of similar reports in the United States and from around the world.

There were several videos on YouTube that showed exactly what we saw. A string of bright orange orbs moving along the same course.

The only explanation I could think of was Chinese sky lanterns released together and caught up in the wind. I reviewed videos of such lanterns to see if they looked similar to what we saw.

The lanterns were, however, quite different. Our objects were much larger, perfectly spherical, and much brighter than any candle-powered lantern. So, perhaps they were UFOs after all.

Area 51

It was May 14th, 1989, in Las Vegas, Nevada. A local television, investigative reporter, George Knapp, conducted an on-camera interview with a person named, "Dennis". Since Dennis was a whistleblower, he wanted to remain anonymous, so he was filmed in the shadows.

Dennis' story was fantastic. He claimed that he had worked at a secret military base, northwest of Las Vegas, that he called Area 51. No one had ever heard of Area 51 up to that time.

He said he worked on a remote part of the base, Sector 4 (S-4) which was located several miles south of the main base at Groom Lake. Dennis *claimed that he was employed as a physicist to reverse engineer a flying saucer that was extraterrestrial in origin.*

Each day, Dennis was flown into Area 51 from McCarron Field in Las Vegas. The carrier was Janet Airlines, a private contractor owned by a company named E.G. & G. This was the private contractor who managed Area 51 and the Nevada Test Site. Then he was taken by bus to S-4.

Dennis claimed that most of S-4 was built into a mountain side and consisted of several, concealed, aircraft hangars.

In a later interview by George Knapp that took place the following November, Dennis appeared under his own name, Robert (Bob) Lazar.

He claimed that his job interview for work at the S-4 facility was conducted by contractor EG&G[CXXI] at their Special Projects

---

[CXXI] **E.G.& G.** at the time was a major government contractor engaged in the nuclear weapons development program. They had a contract to manage the nuclear weapons testing site northwest of Las Vegas. They also managed Area 51, that was adjacent to the testing site.

Headquarters in Las Vegas. His ultimate employer, however, was the United States Navy. His story created a sensation that continues to this day.

Most people have now heard of Area 51. *The legend created by Lazar is that this is the place where the Federal Government takes retrieved Alien technology to reverse engineer, and Alien bodies to be studied and dissected.* It has been glorified in many movies, such as **Independence Day**, and numerous TV shows, like **The X-Files**.

One part of Lazar's story that persists is that captured UFOs are being reverse engineered there. That is, the Alien technology is being investigated to determine exactly how it works. The objective would be to use that knowledge to create a similar craft for human use. Some people have claimed to have seen UFOs maneuvering around Area 51, presumably the captured alien craft that are being tested.

Bob Lazar

For many years, the Pentagon denied that there was any kind of base in the area. The truth is, however, that Area 51 is, and has been since the late 1950's, a strongly guarded, Government facility used for the development and testing of top secret, U.S. aerospace technology.

As can be seen in the Figure below, the location of Area 51 is about one hundred miles northwest of Las Vegas. It is part of the vast amount of barren, moonscape like, federal land in that part of

Nevada. On the maps, the government owned land is divided into numbered Areas.

Formerly, the most well-known Areas to the public were numbers 9 to 15. These were used as the Nuclear Test Site where, *up to the 1960's, nuclear weapons were detonated in deep, underground bunkers.*

As can be seen in the Figure below, Area 51 is adjacent to the Test Site, and contains a dry lakebed named Groom Lake. In the early 1950s, the CIA recognized that it needed a way to gather aerial intelligence over the Soviet Union. They wanted a plane that could cruise at 70,000 feet. This would be out of reach, they thought, of Soviet defensive missiles.

The Agency secretly commissioned *Lockheed's, now famous, Skunk Works in Palmdale, California,* to design and build such a craft. The Skunk Works managers were given the responsibility for finding a suitable place on government owned land to test their design.

They, not the CIA, selected Groom Lake in Area 51 for its isolation and topographical suitability to conduct the tests. The U2's inaugural flight was in August 1957. *This was the first mission at, what was to become, the Groom Lake base.*

The second major project at Groom Lake was the spy plane successor to the U2. The U2 was relatively slow, and, as it turned out, not safe from Soviet missile technology, even at 70,000 feet. Francis Gary Powers, a U2 pilot, was shot down and captured by the Soviets.

The Lockheed Skunk Works was, once again, secretly commissioned to develop a solution. They came up with the ultra-sleek, A-12 Blackbird. A later, better known model with the same "Blackbird" name, was designated the SR-71.

SR-71 Blackbird

The super-fast, high-flying plane took over the Soviet spy flights in 1966. The A-12 flew too high, 95,000 feet, and too fast, over Mach 3, or more than 2200 miles per hour, for the Soviet missile systems. So, no Blackbird was ever shot down. Later, however, the capabilities of advanced, satellite based, spy technology obviated the need for these aircraft, so, after 1999, most of them were relegated to museums.

As the Groom Lake base in Area 51 grew with other advanced development projects, the CIA did not have a large enough staff or the ability to operate the site physically or administratively.

Instead, they contracted a Boston area company to take responsibility for the base. That company was Edgerton, Germeshausen and Grier, Inc. (E.G.&G.).

E.G.&G. was formed after World War II by a group of scientists and engineers from M.I.T. that were heavily involved in the Manhattan project. The firm became a major contractor to the Government for the nuclear weapons program.

The CIA chose E.G.&G. for Area 51 because they were already running the Nevada Test Site that was adjacent to Area 51, and most of their employees were technologically sophisticated and had Top Secret clearances. E.G&G set up their Special Projects Division, based in Las Vegas, to manage the CIA contract.

My second job as a young physicist was with E.G.&G. My responsibilities involved working in laboratories in the Boston area where we simulated the effects of a nuclear explosion on various materials. The goal was to improve the design of the warheads on our ICBMs, to make them more impervious to enemy interference. We used a range of technologies, including the pulsed electron beam accelerator on which I worked.

I never visited Area 51, but I did get to the Special Projects Division Headquarters, and then the Nevada Test Site that bordered on Area 51. I had an opportunity to visit the Test Site as some of my colleagues were spending a few months there preparing for an underground nuclear test.

When I arrived in Las Vegas, I checked in with the Special Projects Division headquarters which managed Area 51. This was before the time when Area 51 became known for the popular stories about Alien technologies, so I was not particularly curious about the place. Furthermore, although I had a Top-Secret "Q" clearance, I did not have a need to know what was going on at the base. So, I did not ask, and they did not tell.

But, in the last few years, many people who worked there have come out about what they did. One was Jules Kabat who was an E.G.&G. Special Projects engineer, employed at Area 51 at about the time of my visit. Many years later he wrote about his two years at the secret base, apparently without violating his oath of confidentiality.

Kabat worked on the development of the A-12 Blackbird. His job, along with his colleagues, was to monitor the radar signatures of the Blackbirds as new materials and design modifications were made to improve their stealth capabilities. His office faced the main runway at Groom Lake, the official name of the Area 51 facility, so he would watch the Blackbirds impressive take offs and landings. He did not report on seeing any signs of "alien technology" during his two years at Area 51.

Several other engineers and pilots also discussed their jobs at Groom Lake. Their activities were related to the major developments in stealth aircraft. The F-117 stealth fighter was born here, as was the B-2 bomber.

I saw a B-2 flying over Boston during one of the Boston Pops, Esplanade concerts on a Fourth of July. It was easy to understand how people reacted if they saw it during its development at Groom Lake. It is a large, black wing, flying low and quietly. It was unlike any other aircraft I had ever seen. It would have been easily mistaken for a UFO if you did not know better.

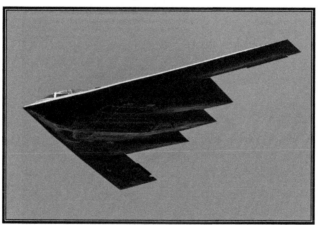

The B-2 Stealth Bomber

The talk of reverse engineering Alien technology may have also had its genesis in what actually happened at Groom Lake. Captured Soviet aircraft were brought there for reverse engineering and testing. It was easy for conspiracy theorists to twist a leak of this type of activity at Area 51 to make it their own.

So, to this day, Area 51 is still an active Air Force base. In addition to the development of the stealth aircraft, the first lethal drones were tested there back in the mid-sixties. More recent projects probably included the stealth helicopters that were used in the raid on Osama Bin-Laden's home in Pakistan.

So, what about Bob Lazar and his story of reverse engineering Alien technology? Over the years, from 1989 to the present, Lazar has stuck to his story about Alien technology at Area 51. However, he has no evidence to support his core claim. He and his tale have been analyzed and rejected by skeptics, and even some Ufologists such as Stanton Friedman.

Universities from which Lazar claims to hold degrees show no record of him ever even having attended, and certainly not receiving any degrees. His supposed former workplaces, such as the Los Alamos National Laboratory, have formally disavowed him as having any professional involvement. However, there is some evidence that has come out more recently that supports Lazar's claim. A well-known, Las Vegas investigative reporter, George Knapp, accompanied Lazar on a visit to the Los Alamos Laboratory. According to Knapp, Lazar knew his way around the lab and was greeted by current employees.

His personal life certainly raises some questions about his character. In 1990, he was convicted for his interest in a prostitution ring and again in 2006 for selling illegal chemicals.

I must say that I am among the skeptics about Lazar and his story. If I were in charge of E.G.&G.'s Special Projects Division and I wanted to staff a super-secret, super-important project like the reverse engineering of an Alien flying saucer, *I would hire a team of fifteen to twenty of the finest technical minds in the country.* All with advanced degrees in physics and aerospace engineering. I would also include a group of top-notch technical and administrative people to support them.

Anyone from E.G.& G. looking to put together such a top-level team would start by looking at their own staff which, at the time, employed hundreds of high level, physicists, and engineers, all with top secret, Q clearances in place. These clearances require an extensive background investigation by the F.B.I. and take months to complete.

I certainly would not hire a non-degreed person with a highly questionable background to take the responsibility for such a job. It is inconceivable that anyone would do such a thing.

On the other hand, I did have an opportunity to listen to Lazar being questioned by Joe Rogan, a very tough interviewer, over a two-hour period in 2019. My impression was more favorable of Lazar. He provided many more details than I had ever heard before. And what he said sounded credible and scientifically reasonable. I was much more impressed with his knowledge of physics for a non-degreed person. This interview increased his credibility in my view, but I still remain skeptical about his narrative.

Lazar is now joined by several other former employees of Area 51 who also claim the existence of Alien technology. Many of these people believe that there have indeed been flying saucer crashes with recovered technology, as well as the bodies of the occupants, some being alive. *The 1947 Roswell incident is one example.*

The U.S. government continues to deny such stories. However, it must be said, *if there were such incidents, Area 51 would be a very logical place to bring the recoveries.* So, we may have to wait until the Government decides to finally declassify all of their relevant files to find the total truth.

The Pentagon 2021 UAP Report

On June 25, 2021, the Pentagon released their Congressionally mandated report on *Unidentified Aerial Phenomena* (UAP).[298] We have always called these objects UFOs or Flying Saucers, but the military changed the terminology.

The principal reason for the change was that, up until this point, anyone who talked about seeing such a phenomenon was mocked as an eccentric or incompetent person. *The associated stigma resulted in very few professional people, such as pilots or radar operators, being willing to report their sightings.*

All of that started to change in 2017, when the first UAP videos were released. One involved Commander David Fravor, now a retired Navy, fighter squadron commander who was based on the Nimitz Aircraft Carrier. He had a close encounter in 2004 off the coast of California.

He, and his wing mate, Lieutenant Alex Dietrich, a female pilot, saw what she called a "tic-tac" shaped object. It was about the size of her F/A 18 fighter, which is fifty-six feet in length. She has said, on the record, that she is not aware of any earth-based technology that could duplicate the flight performance she observed. Many people thought, at the time, that the objects were tests of a secret U.S. project.

Now we know that many more persons in the Military, using highly sophisticated technologies, have been seeing and recording these objects for years. As a result, Congress authorized a Pentagon program, *the Advanced Aerospace Threat Identification Program* (AATIP), which studied UAPs from 2007 to 2012. AATIP was succeeded by the *Unidentified Aerial Phenomenon Task Force (UAPTF)* which presumably continues to this day.

After the unauthorized release of the Nimitz and other videos, Congress passed a law that required the Pentagon to produce a report on all reported military sightings since 2005. That law resulted in the 2021 report, "**Preliminary Assessment: Unidentified Aerial Phenomena**".

The most crucial point about this document is that, after the long history of trying to debunk all UFO reports, the U.S. military has finally confirmed that **UAPs (UFOs) are real** *and very mysterious.*

In addition, another important statement was that *the UAPs are not ours.*

"Gimbal" UAP recorded by Navy Fighter

There are a total of 144 cases of UAPs considered in the Pentagon disclosure. Of those, 80 consisted of incidents that involved observations with multiple sensors. Following is some of the remarkably interesting characteristics of the UAPs that have been recorded by the Navy and others:

"In 18 incidents, described in 21 reports, observers reported unusual UAP movement patterns or flight characteristics. Some UAP appeared to remain stationary in winds aloft, move against the wind, maneuver abruptly, or move at considerable speed, without discernable means of propulsion. In a small number of cases, military aircraft systems processed radio frequency (RF) energy associated with UAP sightings."

The Navy sightings included a wide range of *non-aerodynamic shapes* for the UAP objects. In addition to the tic tac shape, another Navy video shows a rotating "gimbal" like craft as displayed in the Figure above. A third video shows a sphere that disappears at high speed into the ocean without being destroyed. Any conventional flying machine that hit the ocean surface head-on and at high speed, would be shattered into many pieces. It is like flying into a concrete wall.

These underwater vehicles have been tracked by Sonar on board nuclear submarines at speeds of about 70 miles per hour. This compares to the maximum speed of a nuclear-powered attack submarine of about 35 mph.

A fourth group of objects that were seen by a Navy Destroyer, were *three-dimensional objects, shaped like pyramids, and flashing like a strobe.* They were observed and photographed hovering above the Destroyer that was operating off of the U.S. coast. This *form of UAP* was, perhaps, the strangest of them all.

A former military intelligence officer, Luis Elizondo, headed the Pentagon's AATIP program from 2010 to 2017. He has publicly discussed that there are five observables common to these craft as determined by the ATTIP investigation. They include:[299]

1. ***Anti-gravity***: *UFOs seem to be able to generate lift, without any visible propellers, wings, or rocket propellant. This may be a form of antigravity that is yet to be discovered by humans.*

2. ***Instant acceleration***: *Many UFOs are observed accelerating extremely quickly, beyond any known*

object. This acceleration creates massive g-forces that would normally crush any ship or its occupants.

3. **Hypersonic speed without signatures**: *UFOs have been seen traveling at several times the speed of sound. Normally, hypersonic speeds should create loud sonic booms, but UFOs seem to travel silently.*

4. **Low observability**: *UFOs seem to have the ability to avoid detection, cloaking themselves from radar and visual instruments. When they are detected, they sometimes disappear without warning.*

5. **Trans-medium travel**: *Some UFOs seem to travel effortlessly between space, air, and water. That means they can withstand a huge range of pressure, from the high pressure of the ocean to the low (almost nonexistent) pressure of space - and they can maneuver through all three mediums.*

In many cases, there were *swarms of the UAPs* in the areas where they operated around the naval vessels. The swarms range in number from ten to fifty or more. And they appear to be under intelligent control.

The Pentagon does not, in the report, rule out the source of the objects as being Russia or China. However, we know that both of these countries are experiencing similar incidents as our Navy, and also have active UAP study programs. They appear to be taking actions that would be inconsistent with having developed the technology themselves.

Furthermore, and more importantly, U.S. experts have gone on the record as saying *there is no way* that either of these countries could produce craft with the observed capabilities. These vehicles are not just one or two generations beyond our knowhow, but *hundreds of years, or more, ahead of us.*

So, although the government has not done so, we can, with very high probability, **rule out Russia and China as the source of the UAPs**.

John Ratcliffe, Director of National Intelligence (2020-2021) said while discussing UAPs[300]:

"And when we talk about sightings, we are talking about objects that have been seen by Navy or Air Force pilots or have been picked up by satellite imagery that frankly engage in actions that are difficult to explain."

He added:

"(They display) Movements that are hard to replicate that we don't have the technology for. Or traveling at speeds that exceed the sound barrier without a sonic boom."

Ratcliffe also stated that *"there are a lot more sightings than have been made public."* So, the additional data may be dribbled out, over time, in "unofficial" leaks.

The Pentagon also said that it *does not know if the UAPs are extraterrestrial in origin.* They do not have any evidence that this is the case. On the other hand, they said that they *cannot rule out the possibility.*

In the Report, while discussing the origin of UAPs, it says, *"...we may require additional scientific knowledge to successfully collect on, analyze and characterize some of them."* What this means is that the government believes that, to understand the nature of these objects, we will need an understanding of science that is well-beyond our current capabilities.

To be complete in our investigation of the origin of the UAPs, it is conceivable that they are *neither* from our civilization *nor* from one having originated on an exoplanet. John Brennan, a former Director of the CIA (2013-2017) wants to leave open that possibility.

Brennan was the Director of the Central Intelligence Agency from 2013 to 2017. After the Pentagon released the videos of the encounters between carrier-based fighters and UAPs, Brennan was interviewed by Tyler Cowan on his podcast. [301]

John Brennan

Brennan opened the door to the idea that the *origin of the UAPs may be a different life form **not necessarily from another planet**.*

> *"I've seen some of those videos from Navy pilots, and I must tell you that they are quite eyebrow-raising when you look at them.*

> *"You really have to approach it with an open mind but get as much data as possible and get as much expertise as possible brought to bear.*

> *"When people talk about it, is there other life besides what's in the States, in the world, the globe? **Life is defined in many different ways.***

> *"I think it's a bit presumptuous and arrogant for us to believe that there's no other form of life anywhere in the entire universe. What that might be is subject to a lot of different views.*

> *"But I think some of the phenomena we're going to be seeing continues to be unexplained and **might, in fact, be some type of phenomenon that is the result of something that we don't yet understand and that could involve some type of activity that some might say constitutes a different form of life (here on Earth)."***

Brennen seems to imply that we should keep our minds open to the possibility that there may be another form of life closer to home. As a former CIA Director, he would have had access to all the data and research that has been done on UFOs for the last seventy years. And his opinion may be based on that knowledge.

As an example, Dr. Michael Masters, a biological anthropologist, suggests that UFOs are time machines from our own future.[302]

> *"The phenomenon may be our own distant descendants coming back through time to study us in their own evolutionary past,"*

But is time travel even physically possible? *Physicists agree that, according to Einstein's Theory of General Relativity, time travel, both forward and backward, is indeed possible.*

Using the Relativity equations, when you have *a spaceship that can travel close to the speed of light,* and you go on an interstellar voyage to an exoplanet, say five light years away. And then, say, you immediately return to Earth. When you arrive back on our planet and you check the clock on board your ship, it is ten years since you left.

On Earth, however, *according to the Relativity equations,* almost seventy-five years have gone by. *So, in effect, you have gone sixty-five years into the future.* This would be an event very much like Rip Van Winkle experienced in that classic tale by Washington Irving.

Einstein's Theory of Relativity also shows that it is be possible to go back in time using a "wormhole" in space-time that can connect one time period to another.[CXXII],[303]

Wormholes are not easy to find or create with our present technology. But, given one thousand, or one million years, future scientists may have solved the problem and, ever since, they have been travelling back in time to observe their ancestors.

[CXXII] A **wormhole** (or Einstein–Rosen wormhole) is a speculative structure *linking present to past points in spacetime,* and is based on a special solution of the Einstein field equations.

This would be one explanation for people who were living two thousand years ago seeing UFO-like craft and recording them in their paintings. [304]

Beyond Relativity, other advances in physics may explain the UAP. The Pentagon UAP programs used parts of their budgets to fund numerous studies of highly speculative, advanced physics ideas. The purpose was to explore issues such as time travel, possible propulsion systems, the origins of the craft, etc. The Defense Intelligence Agency supplied copies of these reports to the Senate.

The cover letter for the reports was addressed to Senator John McCain, head of the Armed Services Committee at the time. It said:

"Based on interest from your staff regarding the Defense Intelligence Agency (DIA)'s role in the Advanced Aviation Threat Identification Program (AATIP) please find attached a list of all products produced under the AATIP contract for DIA to publish."

Following is a list of the subjects of some of these theoretical studies:

- Advanced nuclear propulsion for manned, deep space missions,
- Teleportation physics,
- Traversable wormholes, stargates and negative energy,
- High frequency gravitational wave communications,
- Antigravity for aerospace applications,
- Positron aerospace propulsion,
- Warp drive, dark energy, and the manipulation of extra dimensions,
- Aneutronic fusion propulsion,
- Negative mass propulsion.

Although many of these topics do indeed sound highly speculative, much of the research comes from reputable sources. Of the thirty eight studies, more than half were the product of individuals working at academic institutions, such as the University of Nevada-Las Vegas, University of Nevada-Reno, The Ohio State University, and the University of St. Andrews in Scotland.

As for the future, Bill Nelson, the head of NASA and a former Astronaut, recently announced that he has asked NASA scientists to study the UAP phenomena. *"Are we alone? Personally, I don't think we are,"* Nelson said during an interview on CNN.

In addition, it is almost certain that the Congress will ensure that other urgent investigations are conducted by the military and intelligence communities. Hopefully, these studies will be made public at some point.

The Pentagon also continues its investigations of UAPs. A group was formed in 2020, called the *Unidentified Aerial Phenomena Task Force. It picked up where AATIP left off.* This program is based in the United States Office of Naval Intelligence and, as far as we know, continues to this day.

Bill Nelson, NASA Administrator

On July 30, 2021, about one month after the Pen⁺ Report was issued, Professor Avi Loeb, a long-timₑ Astronomy Department of Harvard Universᵢ *formation of the Galileo Project.* Iᵗ research endeavor to search for ₑ technological artifacts. The idₑ

technology and research methods to bear in the investigation of UFOs and other phenomena that may have originated from an extraterrestrial intelligence.

Professor Loeb is joined in the project by a group of scientists and engineers who represent some of the top academic institutions in the world. They include faculty members from Harvard, Princeton, Yale, UC Berkley, and Cambridge in the UK, as well as Universities in Scotland, Sweden, Switzerland, and Israel.

Also part of the group is two, well-known former, government officials involved in the UAP programs. Luis Elizondo was the Head of the Pentagon's Advanced Aerospace Threat Identification Program (AATIP). The other is Christopher Mellen, a former Assistant Secretary of Defense, who was instrumental in releasing the Navy UAP videos.

Having such a prestigious group of academics and former government officials working on the study of UFOs would have been unthinkable before the Pentagon UAP Report.

Loeb believes that the best way to show the nature of UAPs is to move away from relying on politicians or military personnel. Instead, professional scientists need to employ sophisticated scientific instruments and research methods to examine the phenomena in a completely transparent manner. They will only depend on the data they collect with their own technologies. This has never been done in any past investigation.

The Galileo Project has three principal areas of research. First, it will obtain high-resolution, multi-detector, UAP Images in order to discover their nature. The scientists will search for UAPs with a network of mid-sized, high-resolution telescopes and detector arrays with suitable cameras and computer systems, distributed in select locations. As far as we know neither the U.S. Air Force, nor anyone else, has ever undertaken such an investigation before.

The second aim of the Galileo Project will be to use existing astronomical surveys to discover and monitor interstellar visitors to our Solar System similar to Oumuamua. The scientists also want to collaborate with people like Elon Musk and his SpaceX ny to have a launch-ready, space mission that can go out to stellar visitor in order to image and study it.

Third, they want to search for potential satellites, one meter in diameter or smaller, that they believe may be in orbit around the Earth and that were placed there by an extraterrestrial civilization in order to watch Earth. The scientists believe that they can find such objects, if they exist, using land-based telescopes equipped with advanced AI, fast-filtering methods.

This endeavor is exciting since it is the first attempt by the science community to use our best resources and knowledge to study the UFO phenomena. It is a privately funded, completely transparent effort that will be closely watched by everyone. After more than fifty years, we may finally get answers to the UFO questions.

Summary and Conclusion

So, where does the status of the UAP phenomena leave us? We do know that these objects are real, under intelligent control and very enigmatic. But it is obvious that *the technology on which they are based is definitely not of our world. Their origin is either from an extraterrestrial intelligence, or from another terrestrial intelligence of which we have no knowledge.* We do not know their intentions, and that is very concerning.

Yet, it has been decades since they were first noticed, possibly even centuries, and none of the UAPs have given any indication of being a significant threat to our security. *So, perhaps we do not have to worry about that possibility.*

Most people, when they hear the term "UFO," conjure up visions of *aliens* who have travelled to Earth from some far-off exoplanet. Yet, others believe *the visitors could be "time travelers" from our future Earth.* Or that, what we are actually seeing are *visitations from" another dimension,"* as opposed to from another planet.

The Interdimensional Hypothesis is one that suggests UFO and alien sightings are actually visits by beings from different "dimensions" or realities. These realities exist alongside the one in which we live, and far from being a modern phenomenon, they go back a long way in time.

There are reports in ancient and modern documents that talk about this possibility. They include "doorways" or "portals" or "stargates" to other worlds and dimensions. Most examples speak of "visitors" entering into or leaving our planet through these doorways.

There is also support in contemporary physics for this idea. The Many-Worlds Interpretation (MWI) of quantum mechanics was proposed by Hugh Everett in 1957.[305]

The Copenhagen interpretation, by Hans Bohr, is the standard for quantum physics. It postulates that the Schrödinger Wave Equation, which describes all of the possible quantum states, "*collapses*" and we see the "actual" outcome of the experiment. In this interpretation, no one has an explanation for why the wave just collapsed to supply the result.

Everett preferred to go along with the logical interpretation of Schrodinger's wave equation. When we see the result of a quantum event that had many other probable outcomes contained in the wave equation, the result we see is the one that occurred in our reality. The other possible outcomes of the wave equation also exist, but they occurred in the many worlds which live in parallel to our own space and time.[CXXIII]

It should not be possible to travel between these different worlds, but perhaps future scientists have solved that problem as well.

_____

[CXXIII] See the section, "Quantum Physics and Consciousness"

Interstellar Expansion Technologies

The astronomers and astrobiologists involved with the SETI program are primarily focused on detecting intelligent radio or light signals emanating from an inhabited planet somewhere in our Milky Way galaxy. But these scientists are also looking for other signs of intelligent life in the cosmos.

Some of these scientists are focusing on studying the newly discovered exoplanets for signs of life. One technique is to use optical instruments called spectrum analyzers which can identify the components in the exoplanet's atmosphere. If they spot gases like oxygen and methane, it would be a possible marker that life exists on that planet. This, of course, would not mean that it would be intelligent life, but such a discovery is a necessary precursor.

A more certain sign of intelligent life would be to find such things as *Dyson Spheres* and *von Neumann Probes*. We would know with certainty that these objects are technological artifacts from an extraterrestrial civilization.

Dyson Spheres

Novelist, Olaf Stapledon, in his book, "Star Maker" (1937), described the idea of collecting enormous quantities of energy from a star:

> "...every solar system... surrounded by a gauze of light-traps, which focused the escaping solar energy for intelligent use".[306]

In 1960, Professor Freeman Dyson of Princeton's *Institute for Advanced Study* popularized this concept in a journal article.[307] Dyson speculated that such megastructures would be the logical consequence of the escalating energy needs of an advanced technological civilization.

Professor Freeman Dyson

We know that only a tiny fraction of a star's energy emissions reaches the surface of any orbiting planet. As a civilization grows, it would reach a point where its energy needs could not be met from just the sources available on its home planet.

One solution would be to build a Dyson Swarm, a massive set of orbiting structures which would encircle its star and enable the civilization to harvest far more energy from their "sun" than any planet-based system. A *Dyson Swarm* is composed of a massive number of movable, independent satellites that, as a whole, compose a *Dyson Sphere*.

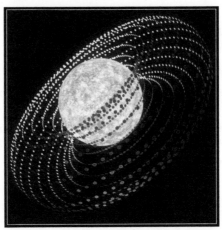

Dyson Swarm

In fact, having access to this enormous amount of energy would be a necessity for an advanced, technological civilization's long-term survival.

So, Dyson proposed that searching for these structures in the galaxy would be a fruitful means to detect such advanced, intelligent, extraterrestrial life.

We discussed Tabby's Star in *an earlier section on exoplanets.* The Kepler telescope typically detected a 1% to 2% drop in the light from a star with exoplanets as the planet passed between Kepler and the star. The amount of decrease would depend on the size of the planet orbiting the star.

In the case of Tabby's Star, the observed decrease of light amounted to almost 22%. This enormous change raised the possibility that it was the result of a Dyson Swarm blocking a considerable fraction of the star light.

Astronomers are considering a range of natural phenomena that are possibilities to explain such a large decrease in the light coming from the star. They include the remnant pieces of a comet in orbit around the star, or an orbiting, massive cloud of dust particles.

If the scientists cannot confirm a natural occurrence as the cause, they *may attribute the decrease to a Dyson Swarm and consider the observation to be a confirmation of an extraterrestrial intelligence.*

Von Neumann Probes

One argument against the idea that UAPs come from a planet outside our solar system is the long distances and travel times involved. But, if an extraterrestrial civilization wanted to spread to other parts of the galaxy, unmanned, *von Neumann probes could be the solution.*

John von Neumann was a well-known, Princeton mathematician. In the middle of the twentieth century, he conceived of a number of creative and interesting concepts. One was the use of automatically, self-replicating machines.[308]

He called such devices *universal constructors*. Other futurists, like Freeman Dyson, expanded his idea to the *use of these mechanisms for space exploration.*

A single, self-replicating probe (SRP) from Earth is sent out into the galaxy. Its first stop is the asteroid belt where it uses mining technologies to obtain the raw materials it needs. It gathers the naturally occurring deposits of iron, nickel, and other minerals found on the asteroids. Then it employs manufacturing technology, like 3D printers, to build a factory to reproduce copies of itself by the hundreds.

The new group of SRPs then set out, each traveling to a different system. When they arrive, the cycle repeats and the SRP creates hundreds of added copies. This group then also sets out on a similar journey.

The original SRP, and all of its copies, carry a package to be left on each planetary object that it visits. The package is designed to establish colonies on suitable planets by spawning biological organisms or using super intelligent AI robots.[309]

The number of probes, having grown exponentially with every landing, would spread out like a giant bubble and, in *less than ten million years*, occupy every corner of the Milky Way galaxy.[310] This is a very short period of time on a cosmological scale.

Given that such an approach is viable for an advanced civilization, it *raises once again the Fermi Paradox*. We should see evidence of those advanced civilizations everywhere we look. But we do not.

In science fiction, the best-known example of a von Neumann probe is found in Stanley Kubrick's 1968 film, "2001: A Space Odyssey." In this movie, a large, black monolith[CXXIV] appears to pre-humans on the African plains and propels them to the next level of evolution.

[CXXIV] **Monolith** – literally a large, single stone. The black monolith appears four times in the film. First in the African wilderness when early man sees it, touches it, and turns into a more dangerous predator. Next in an excavation on the moon. Yet again in orbit around Jupiter. And finally, in the bedroom of the dying astronaut, Dave.

The Black Monolith with Pre-humans

So, the sudden appearance of such a von Neumann probe on Earth would show the existence of an extraterrestrial, intelligent civilization. *In fact, Earth may have been visited recently by just such an object.*

Oumuamua

In October of 2017, an astronomer, Robert Weryk, using the Pan-STARRS telescope at the Haleakala Observatory in Hawaii, noticed a very strange and unusual object during his observations. *He named it "Oumuamua", which, in the Hawaiian language, means a "messenger from afar".*

Weryk and his associates noticed from its trajectory that the object came from interstellar space. That is, it had originated from somewhere outside of our Solar System. This is very unusual. In fact, it had a number of other unusual characteristics that made the astronomers sit up and take notice.

First, the astronomers inferred from its reflection of the sun's light, as the object was spinning along its path, that its shape was more elongated, or flattened, than any known object from our Solar System.

Another intriguing observation involved its trajectory around the Solar System. The astronomers had calculated its expected trajectory using the Sun's gravitational force alone. But when they watched its path, the object deviated from that course. Such a shift could be caused through the "rocket effect" that would be expected from "cometary" outgassing.^{CXXV} *But there was no cometary tail observed around Oumuamua. So, outgassing could not have been the cause of the change in motion.*

These observations caused Abraham (Avi) Loeb, an astronomer with Harvard University, and the head of the Department, to propose in a published paper that Oumuamua could be an alien probe.[311]

Professor Abraham (Avi) Loeb

"We propose that the peculiar acceleration of Oumuamua is caused by the push (not heating) of radiation from the Sun. In order for sunlight to account for the observed acceleration, Oumuamua needs to be less than a millimeter thick but tens of meters in size....

"These inferred dimensions are unusual for rocks, but we do not know whether Oumuamua is a conventional asteroid or comet since we do not have an image of it.

"Our paper [312] shows that a thin object of the required dimensions could survive its journey through the entire Milky Way galaxy unharmed by collisions with atoms or dust particles in interstellar space. We do not know how long its journey had been. If so, what is the origin of Oumuamua?

CXXV **Cometary outgassing**: as a comet orbits about the sun, the surface heating of the object vaporizes ice on the surface and produces the outgassing which acts like a rocket to propel the comet.

"One possibility is that the object is a light-sail floating in interstellar space into which the Solar System ran, like a ship bumping into a buoy on the surface of the ocean.

"A light-sail is a sail pushed forward as it reflects light.... If the reflectivity is high, the inferred size of the object drops from a few hundreds of meters (which was deduced based on the albedo of rock) to a few tens of meters. It is unclear whether Oumuamua might be a defunct technological debris of equipment (from an alien civilization) that is not operational anymore, or whether it is functional. Radio observatories did not detect transmission from it at a power level higher than a tenth of a single cell phone....

Loeb concluded that the scientific community should regularly search for other, similar interstellar objects. He said that we should examine anything that enters the Solar System from interstellar space. This may include NASA, or SpaceX, launching a manned mission to at least take photographs and study the objects composition.

The objective would be to determine with certainty if it is indeed a probe from an extraterrestrial civilization. He has taken an action to have such missions become reality.

On July 30, 2021, about one month after the Pentagon UAP Report was issued, *Professor Loeb, announced the formation of the Galileo Project.* As discussed above, it is a privately funded, research endeavor to seek evidence of extraterrestrial, technological artifacts. It is to be run out of Harvard University.

Given this privately funded, government-independent project, managed, and implemented by a group of prestigious, high-level scientists and engineers, we now have an opportunity to gain a much better understanding of this enigmatic phenomena.

Part V

The Future of

Humankind

V.   The Future of Humankind

Introduction

Our species, Homo sapiens, has only been on the Earth for about 300,000 years of the *3.7 billion years* since the Earth was formed. That is only about one hundredth of 1% of Earth's lifetime. *We are relative newcomers.*

We evolved from a line of early humans that changed as the environment changed. We are the peak result of that evolution to date, but it is not over. Our greatest changes are about to happen, and they are not far into the future.

Up until now, the growth of intelligence on Earth has been the result of the ultra-slow process of evolution. But we are fast approaching the time when gene editing technology will allow humans to increase their intelligence at will.

We are also close to the birth of **machine-based, Super Artificial Intelligence.** *And then the amount of intelligence on Earth will explode to the point where current human intelligence is only a small fraction of the total.*

Ray Kurzweil is a scientist, engineer, futurist and M.I.T. professor. He is now Director of Engineering for Google, a world leader in the development of Artificial Intelligence (AI).

Dr. Ray Kurzweil

Kurzweil's "law of accelerating returns"[313] predicts exponential increases in three major technologies, *genetics, nanotechnology, and AI, that will come together to change our world.* He forecasts that we are reaching a moment in the development of these technologies that he calls *The Singularity.*[CXXVI]

Once *The Singularity* has been reached, Kurzweil says that AI based machine intelligence will be infinitely more powerful than all human intelligence combined. *This will also be the point when, using genetic and nanotechnologies, machine intelligence and humans will merge.*

Then Kurzweil predicts that intelligence will radiate outward from our planet, until it fills the universe. We are already taking the first steps toward that exciting goal as well.

The history of humans has been filled with the drive to explore and expand our population to fill every corner of the world. The stage is now being set, and it is relatively close, for us to begin our expansion out to the stars.

NASA and SpaceX, as well as China, are already planning on a multi-year program to build a permanent, self-sustaining colony on Mars. *The colony could well be thriving by the middle of the twenty-first century* with over 100,000 people living there.

The Singularity could be crossed at about the same time. And super intelligent humankind will be on its way to other stars in our galaxy and ultimately the universe.

To the Stars

Life on the planet Earth has grown and diversified throughout its 3.5 billion years of existence. It appears from scientific observations that there is a biological imperative for life to spread to every corner of the planet. Life forms are even found in the most extreme environments including the bitter cold of the artic

---

[CXXVI] **The technological Singularity**—or simply **the Singularity**-from John von Neumann, is a hypothetical point in time at which technological growth becomes uncontrollable and irreversible, resulting in unforeseeable changes to human civilization.

and Antarctic, and the extreme heat of hydrothermal vents[CXXVII] found in the deepest parts of the ocean.

The dispersion of Homo sapiens from their origin in Africa to every continent on the globe is another example of the biological imperative. And the imperative continues to this day. It has been clear throughout history that we have the need to explore the world and to bring our civilization and culture with us.

It is not surprising, then, that we are showing every sign that this biological imperative will continue, and we will want to spread throughout the solar system, the galaxy, and then the universe.

The Beginning: Building the Mars Colony

If you were over five years old on July 20th, 1969, you probably watched the live broadcast of the first Astronauts to land on the moon. And you have never forgotten the thrill. Hundreds of millions of people from around the world were sitting in rapt attention before their televisions, watching and listening to every word coming from the NASA feed. Since no one had ever done it before, even NASA didn't know exactly what to expect. This made the attempt to land the Lunar Module, named the Eagle, even more dramatic.

Some scientists speculated that the surface of the moon was covered with dust, many feet thick. The previous missions to the surface of the moon by the Surveyor spacecraft found that a manned landing could be made safely. But not all doubt was removed. If the dust were softer and deeper than planned, the Lunar Module's legs might sink into the surface. The craft would either have disappeared or tipped over. And that would have been the end of the mission and probably the end of the Astronauts.

We all held our breaths as the Lunar Module approached the surface. Fortunately, the dust layer was not a problem. Instead, we watched as the Eagle landed safely onto Tranquility Base.

[CXXVII] **Hydrothermal vents** are the result of seawater percolating down through fissures in the ocean crust. These fissures appear where two tectonic plates move away or towards one another. The cold seawater in the area is heated by the hot magma in the fissure and the hot water reemerges to form the vents.

And, later that night, as Neil Armstrong set foot on the Moon's surface and said, *"One small step for (a) man, one giant leap for mankind,"* an unforgettable moment in Earth's history.

The beginning of mankind's journey to populate the universe is the development of a self-sustaining Colony on Mars. Over the last twenty years, a long list of crewed, Mars mission plans have been drawn up by multiple organizations and space agencies from around the globe. *So there is a great deal of worldwide interest in such a project.*

NASA and China's space agency are among the most likely to succeed in launching such a mission. *But perhaps the most likely of all possibilities is a private organization, Elon Musk's* **SpaceX**.

NASAs Plans for Manned Mars Missions

NASA has developed detailed plans for *landing astronauts on the Red Planet by 2039.* It would be done with a stepwise approach that includes a crewed, 2033 trip to the Mars moon, Phobos.

Historically, NASA has been extremely cautious and conservative in its approach to space exploration. Astronauts have died in their programs, but those accidents happened in spite of the most careful, and conservative planning and approaches to the design of the particular missions.

This means that it takes an exceedingly long time for NASA to prepare and complete a mission. And it is very costly as well.

NASA has published its strategy for the human exploration and then prolonged presence on Mars. The concept runs through three distinct phases, leading up to a sustainable human presence on the Red Planet.[314]

The ***first stage*** is "Earth Reliant" and is already underway. The Agency has been using the International Space Station (ISS) to develop and evaluate deep space technologies.

Another major element of this phase is to study and understand the effects of long-duration space missions on the human body.[315] In this regard, numerous astronauts have spent many months on the ISS with careful monitoring of their physical well-being.

The ***second stage*** is called "*Proving Ground.*" The work moves away from reliance on Earth and operates, for most of its efforts, in the space between Earth and the moon's orbit. *This is, called cislunar space.* *The objective is the testing of deep-space habitation facilities, and the validation of capabilities required for human exploration of Mars.*

To conduct this part of the mission, NASA is planning to build *The Gateway.* This is *a lunar orbiting outpost* that will supply support for sustainable, long-term, human return to the lunar surface. It will also *serve as a staging-point for deep space exploration.*

NASA plans to build the Artemis Base Camp on the surface of the moon in about 2024. Gateway and Artemis will allow robots and astronauts to explore the Moon and conduct more science experiments than ever before.

Then, they plan to use what is learned from these missions on and around the Moon *to take the next giant leap: sending the first astronauts to Mars.*

Finally, the **third stage** is the transition to independence from Earth resources. The "Earth Independent" phase includes long term missions on the Martian surface with habitats that only require routine maintenance, and the harvesting of Martian resources for fuel, water, and building materials.

NASA is aiming for human missions to Mars to start in the 2030s. But, true Earth independence, could take decades longer.[316]

An award-winning, 2015 movie, **The Martian**, realistically highlighted the technologies that will be necessary to build a successful Colony on Mars. They include the *ability to make oxygen from the Martian atmosphere,* to *grow plants for food* in a non-Earth environment and to *recover water from the Martian surface.*

Yet *another innovative technology* for a Mars Mission is the development of *deep space propulsion systems* to get the Astronauts to Mars and back to Earth in a timely manner.

The chemical rockets, with which we are all familiar, may not be ideal for the entirety of the Mars mission. Yes, they will certainly be used to get the spacecraft into earth orbit. However, the spacecraft, with the required payload and carrying the most energetic chemical propellant available, would still *not be able to carry enough of that propellant to make the trip to Mars and back in a timely manner.* So, another approach is necessary. One *option that is being considered is a new type of propulsion system.*

In the movie, "The Martian", *an ion thrust system was used to go to Mars from earth orbit.* This is exactly the type of system that NASA plans to use for their real-life mission in the 2030s.

An ion engine uses electric power to ionize a gas. This is the breakdown of the gas atoms into positive and negative components. Then, a strong electric field is employed to accelerate the positive particles (ions) out of the engine at super high speed. The result is a miniscule, but positive, thrust. In zero gravity space, however, even this tiny thrust can accelerate the space craft. The acceleration will be very slow. But it will be continuous, and over time, the craft can reach a remarkably high speed.

And that is the secret to the success of ion propulsion. It can operate continuously for months, or even years, with truly little consumption of fuel compared to a chemical rocket.

The first NASA mission to use an ion engine was the space probe, Deep Space 1, launched in 1998. A more recent application was the ion propulsion system for the Dawn spacecraft, unveiled in 2007. Dawn was used to study the dwarf plant Ceres. Ceres is found in the asteroid belt between the orbits of Mars and Jupiter.

With ion thrusters, the travel time for a crewed mission to Mars can be reduced from six or seven months to just three months. NASA believes that this time saving is critical because it reduces the astronauts' exposure to the hazards of long travel times in space.

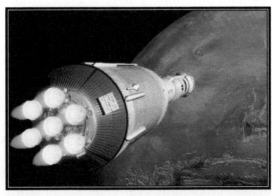

Artist's conception of NASA's X3 Ion Thrusters for Mars Mission

A key to the use of ion engines, however, is to have adequate electric power to operate them. To generate the electric power, NASA has some options. The first that comes to mind is solar power panels. But solar power in deep space and on the surface of Mars has a significant drawback.

NASA InSight Mars Lander with Solar Panels

The inhibiting factor for solar power on the surface of Mars is the distance from the sun to Mars. *The energy available from the sun is inversely proportional to the square of the distance from the sun to the object.*

Given that the Earth is 93 million miles from the sun, and Mars is 144 million miles, on average, the production of electricity by solar cells on Mars surface is only a fraction of the amount they generate in earth orbit. Therefore, *to get the same amount of power in Mars orbit as in Earth orbit, the area of the solar panels would have to increase dramatically.*

Solar cell technology development programs are working toward *making the cells much more efficient.* So, even with the smaller amount of sun power available on Mars, more of that power would be converted to electricity. If the efficiency improvements are great enough over the next several years, solar panels may be able to play a significant role in a crewed Mars mission.

Another power source used for many past Mars missions is a Radioisotope Thermoelectric Generator (RTG). It is a self-contained device, so it doesn't have the problem associated with the solar panels.

This generator uses a small amount of Plutonium-238, a relatively safe radioactive material. The isotope is naturally decaying and produces heat energy with the decay. The heat is then transformed into electricity using a device called a Thermocouple.

The amount of power produced by these units is adequate to run the ion engine during the flight. Upon landing on Mars, the RTG power sources can then be used to operate ground-based equipment.

The SpaceX Manned Mars Colony

If humankind is to go to Mars and set up a Colony anytime soon, it will *most likely be initiated and undertaken by private corporations* rather than the government. Because of the massive expense, however, governments and other wealthy companies will be necessary to fund and fully implement the plans.

Elon Musk, founder of SpaceX, is almost certainly the person that has the best chance for the first Mars landing. In fact, he is well on his way to carrying out that objective.

He hopes that the first human landings on Mars by SpaceX could take place as early as 2026. That is *almost fifteen years sooner than the projected manned landings on Mars by NASA*[317] *or China.*

Speaking at the *International Astronautical Congress* in Adelaide, Australia in May 2020, *Musk's shared his vision to build the Colony to be self-sustaining at one million residents. And his goal is to reach this stage by 2050.*

Musk unveiled plans to send two unpiloted, cargo missions to scout water sources and build a fuel plant on Mars in 2024. Two years after that, 2026, he plans to send four more flights: two additional, unpiloted cargo missions and two carrying astronauts.

There is a significant difference between SpaceX and NASA in their approaches to a Mars mission. Elon Musk was asked at a press conference if astronauts could be guaranteed of a safe return home.[318]

"No. Your probability of dying on Mars is much higher than Earth. Really, the ad for going to Mars would be like Shackleton's ad for going to the Antarctic [in 1914].

It's gonna be hard. There's a good chance of death, going in a "little can" through deep space. You might land successfully. Once you land successfully, ... there's a good chance you'll die there. We think you can come back; but we're not sure."

Musk has developed a detailed plan to achieve his dream. The *first step is to create cheap, reliable, reusable rocket technology* that can lift into space all of the cargo and humans needed for the mission. He began working on that effort several years ago and is close to achieving that goal.

His *first crewed flight on a SpaceX rocket* took place in May of 2020. The spacecraft, named Endeavour, was on top of a well-proven, SpaceX Falcon 9 Booster. It carried NASA astronauts Douglas Hurley and Robert Behnken to the International Space Station in the first crewed orbital, spaceflight launched from the United States since the final Space Shuttle mission in 2011. *It was the first ever, crewed mission run by a commercial provider.*[319]

The space craft being developed by SpaceX for interplanetary missions is called Starship. The name is taken from the spacecraft of the Star Trek, television series and subsequent movies. *Everyone knows of the Starship Enterprise with Captain James Kirk in command.*

Also in May of 2020, Starship, serial number 15, successfully completed SpaceX's fifth, high-altitude test flight of a Starship prototype from the SpaceX Starbase in Boca Chica, Texas.

Starship Test Rocket

The *Boca Chica launch site* is about twenty miles east of Brownsville, Texas, on the U.S. Gulf Coast. It *is a private spaceport*, test facility and rocket production operation. SpaceX constructed it and manages the location.

The Booster system for Starship was nicknamed, tongue-in-cheek, the BFR, for Big "Freaking" Rocket. Its official name is Super Heavy, and it will be used to allow Starship to escape Earth's deep gravity well on its way to the moon and then Mars.

When stacked with a Starship, the total height of the launch system is close to four hundred feet. That compares to about 363 feet for the Saturn rocket that carried the Apollo astronauts to the moon in the late sixties and early seventies. *A major difference between Saturn and Starship is that all of the major components of Starship land back on Earth and are reusable.*

SpaceX Starship mounted on the Super Heavy Booster

SpaceX expects that, by 2026, the Starship will be in frequent use as a long-duration cargo and passenger carrying spacecraft.[320]

Musk's plan is to first build a crewed base on Mars to set up an extended presence on the planet. Then, the long-range goal, decades into the future, is to grow the base into a self-sufficient Colony the size of a city.[321]

For the journey to Mars, *the Starship will be re-fueled in low Earth orbit.* And, *once it lands on Mars, it would need to be refueled again to make the journey home.*

So, an early mission to Mars is to build a plant to produce the propellant for use in return trips to Earth. The first astronauts will know that they will not be able to return to Earth without the successful construction of the fuel plant.

The plant would use sub-surface ice and atmospheric Carbon Monoxide to make methane and oxygen, the main propellants for the Starship.

As a result, before humans arrive on Mars, there will be *two, robotic cargo flights* to the Mars Alpha Base site. The current *timing for this trip is 2024.* These flights would deliver a large solar panel array, mining equipment, surface vehicles, as well as food and life-support infrastructure.[322]

Mars Alpha Base, as termed by SpaceX, will likely be on the smooth plain of Arcadia Planitia. It will be located at about forty degrees latitude of the planet. Arcadia Planitia combines flat terrain, potential deposits of water ice and an equatorial region well-suited for solar power.[323]

Scientists suspect that there are *massive, subsurface water ice deposits at Arcadia Planitia.* This material will be necessary for drinking water and as a raw material to make the rocket propellant.

Then in *2026, four more Starships* would head to Mars. First, there would be another two robotic cargo ships. Each landed cargo mass will be about one hundred tons of usable payload. They would be *followed by two ships carrying crew members.*

The Starship used for human transport will feature forty passenger cabins, each of which can accommodate *two or three* people. This would be a maximum capacity of about 100 passengers per trip.

For the first few trips, however, there will only be about twelve crew members making the voyage. The reason is that the Starships themselves will be used for the initial habitat for the crew, and the crew will need the extra room for food and supplies.

The job of the first crews will be to build and trouble shoot the propellant production plant to make sure that it is working. They will also deploy the solar farm so the site can have power.

Then the crew will assemble landing pads for the future Starship arrivals. Yet another major task will be to *assemble greenhouses* to be used for the production of food for the settlers.

Artist's conception of the first Mars Colony

Once the basics are taken care of, many more settlers can come to the camp. They would all be volunteers who would agree to stay on Mars for an exceptionally long time and perhaps, never return to Earth.

Habitats would have to be constructed for the population. The raw materials could come from the Red Planet, and 3-D printers would be used for the manufacturing.

Since the fully loaded Starship capacity for humans is one hundred people per trip, it would, therefore, take ten thousand trips, conducted over many years, to bring in the one million people. But, to support these people, even more cargo flights would be needed.

SpaceX estimates that it would take ten cargo trips for every human trip. Therefore, we are talking about over one hundred

thousand trips to build the Colony to the one million resident level.
³²⁴

All of this activity would require an enormous amount of capital. While Musk is pouring his personal wealth into the project, many more billionaires and private corporations would also have to invest in the endeavor.

Therefore, a Colony on Mars would have to promise significant returns to the investors in order to attract the necessary capital. Such returns may be possible with the discovery of substantial amounts of valuable raw materials that could be mined and used on Mars, as well as transported back to Earth.

Terraforming the Red Planet

It is hard to believe, but billions of years ago, Mars was Earth-like. It was warm enough to have liquid water on its surface in the form of lakes, rivers, and oceans. There was a dipole magnetic field similar to ours that shielded it from cosmic radiation, and it had a relatively thick atmosphere. In fact, it was enough like our planet that it could have developed life as we know it and probably did.

Significantly, the Red Planet *lost its magnetic field* sometime between 3 to 4 billion years ago. This allowed the solar wind, an incessant stream of energetic, charged particles coming from the Sun, to strip away most of the planet's atmosphere and surface water, turning Mars into the chilly desert we see today.[325]

Now, the environment of Mars, the fourth planet from the Sun, is worse than the harshest conditions anywhere on planet Earth. The surface is desert-like with no visible plant or animal life. It is dusty and rocky with a very thin atmosphere consisting primarily of carbon dioxide and a little water vapor. It is a dynamic planet, however, with seasons, polar ice caps, mountain ranges and extinct volcanos.

Artist's conception of SpaceX Colony at Arcadia Planitia

Being further away, it takes 687 days to revolve around the Sun, compared to 365 days for Earth. It rotates on its axis every 24.6 hours, which is nearly the same as Earth.

Mars has two tiny satellites, named Deimos and Phobos. They are most likely small asteroids drawn into Mars' gravitational pull. Deimos and Phobos have diameters of just 7 miles and 14 miles, respectively. This compares of a diameter of 2,159 miles for our Moon.

Being about forty-eight million miles further away from the Sun, *the level of light on the surface of Mars is about sixty percent of that found on Earth.* This is one factor that makes Mars a much colder planet than Earth.

Temperatures on Mars *average minus 81° Fahrenheit* (F), so water cannot exist on the surface in a liquid form. The range of temperature around the planet range from *minus* 220° F in the wintertime at the poles, to plus 70° F at the equator. This compares to an average temperature on Earth of *plus* 58° F, with a range of minus 81° F to plus 116° F.[326]

An exacerbating factor that hastened Mars loss of its atmosphere after it lost its magnetic field, is *its much lower gravity compared to Earth.* With a diameter of 4,220 miles, Mars is much smaller than Earth, which has a diameter of 7,926 miles. *As a result, the surface gravity of Mars is only about 38% of that of Earth.*[327]

Earth, left, compared to Mars

To have a colony on Mars that is suitable for humans, it will be necessary to build a habitat that compensates for this harsh environment. That is exactly what is planned by NASA and SpaceX, as described in the earlier section.

But a very long-term *goal for NASA is to "engineer" Mars with the aim of returning it to its previous, Earth-like state. A very tall order indeed.* Yet, there are scientists who are considering what it would take to reach that objective.

To *terraform*CXXVIII Mars into a planet that would be suitable for human habitation will require three major, interrelated changes. First, a magnetosphere would have to be built up. Then a suitable atmosphere would have to be created. Finally, the temperature at the surface would have to be raised.

Estimates are that it could take hundreds, or even thousands, of years to accomplish these steps. But the ultimate goal is to have a second home planet for humankind, and it is worth the investment in time and resources.

Building the magnetosphere is the first step. The reason is that such a field is necessary to protect the atmosphere and any inhabitants from the solar wind.

Dr. James Green, Director of NASA's Planetary Science Division, has proposed a plan to create an artificial magnetic field around the Red Planet.[328]

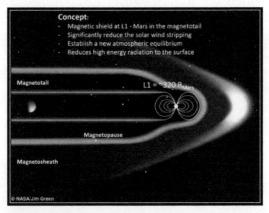

NASA Concept for Mars Magnetosphere

CXXVIII **Terraform** - transform a planet so as to resemble the earth, especially so that it can support human life.

Dr. Green said while discussing the idea:

"In the future it is quite possible that an inflatable structure(s) can generate a magnetic dipole field at a level of perhaps 1 or 2 Tesla (or 10,000 to 20,000 Gauss) as an active shield against the solar wind."

The dipole would be composed of a large electric circuit, powerful enough to generate a planetary, artificial magnetic field. It would be placed in an orbit around Mars *at a "Lagrange Point"*.

These are points in space where objects positioned there *tend to stay put relative to* the planet and the Sun. This is because the gravitational pull of two large masses, *in this case Mars and the Sun*, precisely equals the centripetal force required for a small object to move with them.[329]

As shown in the above figure, the magnetic field generator, placed at the first Lagrange point, would effectively divert the solar wind, and protect Mars. The current, very thin atmosphere would have a chance to begin to increase as natural processes release gases into the atmosphere and they are retained instead of being stripped away.

Once it is clear that an adequate magnetic field is in place, then the process of artificially building the atmosphere can start. The atmospheric pressure on Mars is currently *less than 1% of that on Earth*. So the first step would be to release gases, mostly CO_2 and water vapor, from various deposits around the planet, to thicken the atmosphere. These gases are contained in the Martian polar caps, the soil and in certain minerals that could be mined.

Since the CO_2 and water vapor are "greenhouse" gases they would have the effect of preserving the thermal energy near the surface of the planet. This process would eventually increase the surface temperature to the point where liquid water is stable.

So over the next hundreds of years, if all went according to this plan, we could restore as much as 1/7th the amount of liquid water that Mars once had in its oceans.

Moreover, as the planet heats up, more CO_2 will be released from the frozen reserves on the poles, enhancing the greenhouse effect. *This means that the two processes of building the atmosphere and heating it would augment each other, favoring terraforming.*

But, even then, since Mars has 38% of Earth's gravity, it can only keep an atmosphere of about 0.38 bar. *In other words, even a terraformed Mars would be very cold by Earth standards and its air would be as thin and chilly as the Himalayan mountains.*[330]

But the big question, is there enough easily available CO_2 and water near the surface of Mars to make such a scenario possible?

For many years, NASA spacecraft, such as the Mars Reconnaissance Orbiter, and Mars Odyssey spacecraft, have been making observations on the amount of CO_2 on the planet. They also have data on the rate of loss of the Martian atmosphere to space as observed by NASA's MAVEN (Mars Atmosphere and Volatile Evolution) spacecraft.

Christopher Edwards of Northern Arizona University studied the data and said:[331]

> *"These data have provided substantial, new information on the history of easily vaporized (volatile) materials like CO_2 and H_2O on the planet, the abundance of volatiles locked up on and below the surface, and the loss of gas from the atmosphere to space."*

Bruce Jakosky of the University of Colorado, Boulder, lead author of the study, said:

> *"Our results suggest that there is not enough CO_2 remaining on Mars to provide significant greenhouse warming were the gas to be put into the atmosphere; in addition, most of the CO_2 gas is not accessible and could not be readily mobilized. **As a result, terraforming Mars is not possible using present-day technology.**"*

But we are talking about thousands of years to do the job, so we can assume that the technology will be developed to find and release these materials that may be buried deep on the planet. Then there will be enough greenhouse gases to increase the atmosphere as much as is possible, given Mars' low gravity, and thereby warm the planet as planned to the point where liquid water will be possible.

Another major challenge for terraforming Mars would be to make the atmosphere breathable for humans. We may be able to duplicate what happened on Earth.

It was 2.5 billion years ago that *cyanobacteria, through photosynthesis, converted our atmosphere of methane, ammonia, and other gases, into the oxygen-rich one of today.*

But there is a complicating problem for using this process on Mars. It receives less than half the sunlight as Earth and has a global dust storm problem that makes visibility worse.

Researchers have suggested, therefore, that we introduce special microorganisms on Mars *that photosynthesize in low light to create breathable air for humans.* When paired with other organisms, an entire life cycle could be created on Mars with the result being a favorable blend of gases.

The vision for Mars is to make it a second home for humankind. Even if it takes thousands of years to do so. Then we can be more assured that humans will be preserved in the universe. We would not have to worry that *an Extinction Level Event* on Earth, such as a major comet or asteroid impacting the world, would eliminate the human race. Our siblings on Mars would survive, along with our civilization, to ensure that consciousness continues in the cosmos.

From Homo Sapiens to Homo Machina

The Evolution of Man

To know where humankind is headed in terms of our evolution, it is useful to start with an understanding of how we reached our current state as Homo sapiens. Darwin's evolution by random mutations and natural selection is a "magical", long, and terribly slow process. It started on Earth with single cell creatures *about 3.6 billion years* ago and has resulted in the myriad of complex, sophisticated life forms we see today, culminating in Homo sapiens.

Natural selection, as outlined in Darwin's 1859 book, **On the Origin of Species**,[332] occurs when a genetic mutation, say a larger brain with a more sophisticated structure of neurons and synaptic connections, is passed down through generations, because it affords fitness benefits to its carriers. As a result, these *more fit creatures are able to survive and pass along their genes. Eventually the mutation becomes the norm.*

The history of the development of Homo sapiens, through the process of evolution, is still hazy. Paleoanthropologists have been putting the pieces together for almost two hundred years, but, because of the incomplete fossil record, there are still many gaps and unanswered questions.

The following is what we believe is the *most likely, general path of the evolution of Man from five million years ago up to the present day It is based on the findings of ancient, hominin fossils over the last two hundred years. It includes all of the major hominins but leaves out a few of the lesser-known specimens like Homo ergaster and Homo rudolfensis.*

The *discoveries* of the various hominins[CXXIX] were in a random order of time, but they are *presented* here in their *evolutionary order*, from the oldest creature to the latest, us.

[CXXIX] **Hominin** – the group consisting of modern humans, Homo sapiens, extinct human species and all our immediate ancestors (including members of genera Ardipithecus, Paranthropus, Australopithecus, and Homo.

Ardipithecus ramidus

The first hominin was Ardipithecus ramidus. He separated from a common primate ancestor with chimpanzees about four and a half million years ago. The Ardipithecus fossils were discovered between 1992 and 1994 by American paleoanthropologist, Tim White, a professor from U.C. Berkley.

Professor Tim White

White found the fossils in the Middle Awash area of Ethiopia. This section is along the Awash River in Ethiopia's Afar Depression.

The name "ramidus" came from the word "ramid" in the Afar language of Ethiopia. It means "root" and refers to the idea that this new species was close to the root of humanity.

As can be noted from the Figure below, the creature was close in appearance and form to the current day chimpanzee. Chimpanzees are our closest relative in the family of Great Apes. They share about 98.5% of our genomic identity.[333]

Ardipithecus ramidus

Australopethicus afarensis

Next up in the evolutionary procession of hominins was *the genus, Australopethicus.* It was in 1974 that paleoanthropologist, Donald Johanson, of the Cleveland Museum of Natural History, was working at another site in the Lower Awash Valley near Hader, Ethiopia, when he discovered a pre-historic skeleton. It turned out to be the first of another new genus. The specimen *was a female hominin of the species Australopithecus afarensis, which he nicknamed, Lucy.*[334]

Australopithecus Afarensis

Australopithecus hominins were ape-like creatures that had teeth like humans, but larger in size. All of the members could walk upright on two legs. Compared to later hominins, their skulls were smaller, accompanied by smaller brains.

Dr. Donald Johanson

A 2016 study proposes that Australopithecus afarensis was, to a large extent, a tree-dwelling creature. The exact amount, however, is debated.[335]

This genus consisted of three other significant member species including Ramidus, Africanus and Robustus. They lived primarily on the African mainland. Their genus was relatively long lived and was in existence for 700 thousand years, from 3.7 to 3 million years ago.

Paranthropus boisei

Starting in 1951, the famous paleoanthropologists, Louis Leakey, and his wife, Mary, worked in the *Olduvai Gorge in Tanzania, Africa, seeking hominin fossils.* Their efforts were important in showing that, indeed, the known line of hominins evolved in Africa.

Louis and Mary Leakey

It was in 1959 that Mary Leakey made a major discovery when she found the "Zinj" skull. She and her husband concluded that it was *a new species and named it Paranthropus boisei.* This discovery was one of the most famous finds from Olduvai Gorge.

Paranthropus lived in environments that were dominated by grasslands, but also included more closed, wet habitats associated with rivers and lakes. The species was particularly characterized by its specialized skull that was adapted for heavy chewing. He had large temporalis muscles from the top and side of the skull down to the lower chin. This powered the creature's massive jaw.

Paranthropus boisei

Paranthropus had exceptionally large, cheek teeth, four times the size of those of a modern human. He also had a flat, relatively big-brained skull, and the thickest dental enamel of any known, early human.

Homo habilis

The Leakeys continued their work at Olduvai Gorge between 1960 and 1963. It was then that they made another major discovery. They uncovered the fossilized remains of a unique hominin. Its features were quite different from those of Australopithecus. So, it was declared the first member of the genus, Homo. *They called it Homo habilis.*[336]

"Habilis" means "handyman" in Latin. *That is because habilis was found at a site with a myriad of stone tools*, so he was thought to be the first hominin to use tools to a significant extent.[337]

Homo habilis' face was similar to Australopithecus, but with a slightly larger brain. The skull and brain size show that he may have been capable of speech. He was around 5 feet tall and walked erect.

These creatures lived about 2.6 million years ago. They were good hunters but also ate scavenged meat. Their primary use of tools was for butchering the animals they caught or found.

Homo Habilis

Homo erectus

Following Homo habilis came Homo erectus, or "upright man". The earliest occurrence was about two million years ago, so they are contemporary with Australopithecus.[338]

The first Homo erectus fossils were found in 1890 on the island of Java, which is now part of Indonesia. The discoverer was a Dutch army surgeon named Eugene Dubois.

Dr. Eugene Dubois

Homo erectus were cave dwellers and the first human ancestor to spread throughout Eurasia. Their continental range extended from the Iberian Peninsula to Java.

They had a relatively large brain size. The Java Man and Peking Man, examples of Homo erectus, had brain capacities similar to modern humans at 1300 cubic centimeters. This does not mean, however, that this hominin's brain power was equivalent to Homo sapiens.

Homo erectus had a humanlike appearance, body proportions and gait. It was the first human species to have exhibited a flat face, prominent nose, and possibly sparse body hair coverage.

Analysis of their vocal capabilities indicated that they may have had the ability for speech. He certainly knew how to make and use tools. He was carnivorous and was the first hominin to employ fire for cooking.[339]

Homo erectus

The cooking of food may have been a key to larger brain development. Brains are only 2% of the body's weight but require about 25% of the body's caloric needs. Cooking the food provided Homo erectus with more densely packed calories in the food they consumed. As a result, they could support a much larger brain for their smaller sized body. Evolution did the rest.

Homo heidelbergensis

An African form of Homo erectus evolved into Homo heidelbergensis about 600 thousand years ago. The first fossil of this hominin was discovered in 1908 in the village of Mauer, near Heidelberg, Germany. A laborer, working in a sandpit, inadvertently uncovered part of the fossilized remains that turned out to be Homo heidelbergensis.

Later, German scientist, Otto Schoentensack was the first to describe the specimen and proposed the species be named Homo heidelbergensis. This hominin had an exceptionally large ridge on its brow, and a larger braincase and flatter face than older, earlier human species.

Dr. Otto Schoentensack

Heidelbergensis was the first, early human species to live in the colder climate areas of Europe and Asia. They had short, wide bodies that were an adaptation for conserving heat. And they had an overall, robust build.

It was a time when these hominins had definite control of fire and used wooden spears for hunting large animals. Heidelbergensis *also broke new ground; it was the first species to build shelters, creating simple dwellings out of wood and rock.*

Their brain size was about 1,200 cubic centimeters, slightly smaller than a modern human. They ranged in height from 5 feet 2 inches to 5 feet 5 inches for females, and five feet seven inches to five feet eleven inches for males.

Homo neanderthalis

The first modern human fossils were found in Engis, Belgium in 1829 by Dr. Phillipe Charles Schmerling, a medical doctor. He is considered to be the "father" of paleontology. At the time, he believed the specimen to be Homo sapiens.

Then, twenty-seven years later, in 1856, another set of similar fossils were discovered. They were found by some limestone miners who were working in the Feldhofer Cave located in the Neander Valley. This is small valley of the river Düssel in the German state of North Rhine-Westphalia. These fossils were also thought to be Homo sapiens

Homo heidelbergensis

It took until 1864 for geologist, William King, of Queen's College, Galway, Ireland, to recognize that both sets of fossils were a distinct species from Homo sapiens.

At a meeting of the *British Association* in 1863, he made his case for the new species and proposed the name of Homo neanderthalis. The written version of his presentation was published in 1864.[340]

Professor William King

Dating the fossil finds revealed that Neanderthals appeared about 200,000 years ago. They looked similar to modern humans, but they were shorter, stockier with prominent brow ridges and wide noses. Their brain size was larger than ours and *they may have had the potential to be even more intelligent than we are.*

If their brain structure was similar to ours, they had more neurons in their neocortex which could have made them smarter. But, given the great technological strides made later by Homo sapiens, it *may have been that the Neanderthal's brain structure was not the same,* and that Homo sapiens developed a smaller, but much more powerful brain, with many more possible, synaptic connections.

In any case, Neanderthals were more accomplished than earlier hominins. An incredibly significant achievement of Neanderthals was that they hunted in groups. This means that they could successfully communicate, coordinate, and cooperate together. In addition, they employed sophisticated tools to a significant extent and buried their dead.[341]

Yet another significant, Neanderthal accomplishment was their wall paintings found in Spanish caves.[342] They did the work about sixty-five thousand years ago, making the paintings the oldest cave art ever discovered.

Neanderthal Cave Art

The Neanderthals did many diverse types of paintings including human figures and various animals. They also created the images shown above by placing a hand on the wall and then blowing red or white pigments at it to create fascinating compositions.

Some scientists believe that the Neanderthal's achievements in cave art is the moment when the modern, human mind was born. They think that the Neanderthals were the first hominin to imagine and dream as modern humans do today.

Neanderthals went extinct about thirty thousand years ago. No one knows why, but it may be that they were assimilated into the modern human genome.

In fact, Neanderthal-inherited genetic material is found in all non-African populations. It is estimated that 1.5 to 2.1 percent of the DNA of modern humans originated with the Neanderthals.

Neanderthal group living in a cave

Homo sapiens

Hominins reached their peak with Homo sapiens. It was during a time of dramatic climate change 300,000 years ago that Homo sapiens evolved in Africa. About 120,000 years ago, the subspecies, *Homo sapiens sapiens, modern people*, evolved in Africa and migrated to the North. First to the Middle East, then Europe and later Asia, Australia, and the Americas.

It is clear to scientists that *changing environmental conditions stimulated all of the major developments in human evolution* over the last five million years. Only the smartest and strongest hominins survived and went on to reproduce.

The points in time of changing environmental conditions *are noted on the chart below.* Scientists at the Smithsonian have found that *there is a correlation of climate fluctuations with the record of changes of oxygen isotopes with time.*[343]

In turn, the oxygen isotope information can be understood by measuring the amount of oxygen in the microscopic skeletons of foraminifera[CXXX] that lived on the sea floor at the time of the climatic change.

Icons marking evolutionary changes are shown on the base of the graph, with *time **flowing right to left**. They are*: (a) hominin origins, (b) habitual bipedality, (c) first stone toolmaking and eating meat/marrow from large animals, (d) onset of long-endurance mobility, (e) onset of rapid brain enlargement, (f) expansion of symbolic expression, innovation, and cultural diversity.

The oxygen isotope amounts are an indicator of changing environmental conditions. As can be seen in the Figure below, there are two main trends: 1. *an **overall decrease in temperature*** and 2. ***a larger degree of climate fluctuation** over time.*

[CXXX] **Foraminifera** - a single-celled planktonic animal with a perforated chalky shell through which slender protrusions of protoplasm extend. Most kinds are marine, and when they die, their shells form thick ocean-floor sediments.

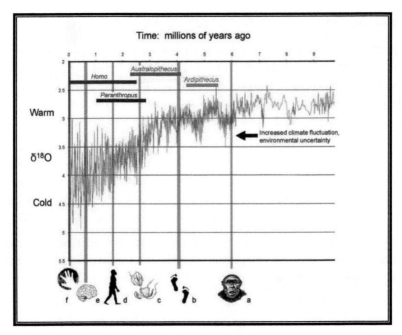

Smithsonian Institution-Oxygen isotope curve, over the past 10 million years.

As can be noted, the amount of variability in environmental conditions *was greater in the later stages of human evolution. This was especially true during the development of Homo sapiens.*

In the graphic, the amplitude of oscillations grew during the past 2.5 million years. *The evolution of the genus Homo and the significant adaptations of Homo sapiens were associated with the largest oscillations in global climate over the last one million years.*

One way that organisms, such as a member of the Homo genus, *cope with environmental fluctuation is through genetic adaptation.* The organisms under stress may have several **alleles**, *or different versions of genes*, in the population, and at different frequencies. As conditions change, Natural Selection favors one allele, or genetic variant, over another and **organisms with that allele survive and reproduce.**

Genes that can enable a range of different forms under different environments (*phenotypic plasticity*^{CXXXI}) *can also help an organism adapt to changing environmental conditions.*

Like other early humans, erectus and habilis, that were living at this time, *Homo sapiens gathered and hunted food, and evolved behaviors that helped them respond better to the challenges of survival in the unstable environment.*

For example, they needed shelter and clothing to help them cope with the weather conditions as temperatures declined. And they had to be clever enough to find the necessary shelter, food, and animal skins to survive under those challenging conditions.

The evolutionary changes that took place at this time helped sapiens to stay alive and reproduce, while Homo erectus and Homo habilis did not and became extinct.

Anatomically, modern humans can generally be characterized by the lighter build of their skeletons compared to earlier humans. Modern humans have large brains, which vary in size from population to population and between males and females. The average size is approximately 1300 cubic centimeters. This is about the same size, *or actually smaller,* than their predecessors Homo erectus and Homo Neanderthalis.

Yet, Homo sapiens showed significantly increased brain power compared to these earlier Homo species. Perhaps, the *brain's structure* (neuron density in the neocortex) *became more complex with broader, stronger synaptic connections, resulting in increased intelligence in Homo sapiens.*

One sign of the increased brain power of Homo sapiens is the sophistication of their tool making. Their tools consisted of a variety of smaller, more complex, refined, and specialized tools including composite stone tools, fishhooks, and harpoons, bows and arrows, spear throwers and sewing needles. [344]

---

^{CXXXI} **Phenotypic plasticity** - refers to the changes in an organism's behavior, morphology, and physiology in response to a unique environment.

Housing this big, complex brain involved the reorganization of the skull into what is thought of as a "modern" structure. It is a thin-walled, high vaulted skull with a flat and near vertical forehead. Modern human faces also show much less, if any, of the heavy brow ridges and protruding lower jaw with the large teeth of other early humans.

The complex brains of modern humans enabled them to interact with each other and with their surroundings in new and diverse ways. They ate a variety of animal and plant foods *which they cooked,* thereby allowing them to receive more densely packed calories, and, which, in turn, *allowed for the development of a larger brain.*

They lived in shelters; they built broad social networks, sometimes including people they had never even met; they exchanged resources over wide areas; and they created art, music, personal adornment, rituals, and a complex, symbolic world.[345]

The first modern humans began to move outside of Africa about seventy to one hundred thousand years ago. It was not until about fifty thousand years ago that they started to develop language. *This is another sign that brain power increased while the size of the brain did not.*

The diagram below summarizes the evolution of hominins leading to Homo sapiens. Only chimpanzees and humans survive to this day. The split with a common ancestor with chimpanzees happened about five million years ago with Ardipithecus being our earliest, hominin relation. Then our line traces back to the Australopithecus, Homo habilis, Homo erectus and Homo neanderthalis genera.

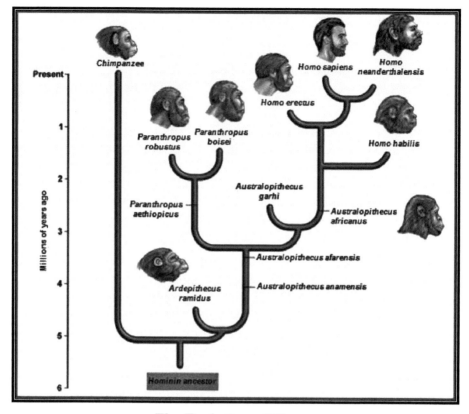

The Evolution of Man

Homo sapiens is the peak of the evolutionary process *to date. But it is to be determined whether we humans survive whatever fast-changing environmental conditions that we encounter by evolving quickly enough into something superior; or we become extinct like so many hominins did before us.* **The odds in our favor have, however, increased since we no longer have to depend on Darwinian evolution.**

The End of Darwinian Evolution

The Genetic Revolution

Slow, Darwinian evolution has been continuing in Homo sapiens over the last one thousand years. For several reasons, however, it will not continue.

First, with the ease of international transportation over the last one hundred years, human populations have been coming together and intermingling all over the globe. While Natural Selection leads to the *adaptive divergence*[CXXXII] of populations in *different environments*, the exchange of individuals between populations (and their genes) opposes that divergence. Instead, population shifting and mixing leads to homogenization.[346]

As a result, some parts of our populations have been combining genetically and altering the overall appearance and capabilities of our species with a tendency toward becoming more alike. This trend will continue in the general population for some time to come. And will reach equilibrium within the next fifty generations, or about one thousand years.[347]

On the other hand, there have been small groups of humans that have been geographically confined and *remained relatively untouched* by the global population movements. These people include those who live in deep, remote rain forests or isolated Pacific islands. Such classes of people have been found to *continue to evolve by natural selection over generations, based on the environmental and societal pressures exerted on them.*

A good example of this type of contemporary, Darwinian evolution is the Bajau people of Southeast Asia.

The Bajau are nomads who live permanently "at sea". They originated in the Philippines but came to Malaysia. Because of their status as outsiders, they were forbidden from living on

---

[CXXXII] **Adaptive divergence** - Divergence of new forms from a common ancestral form due to adaptation to different environmental conditions.

Malaysian soil. So they were forced to live and make their living offshore. Their housing is in stilt-huts built along coastal waters. *Hundreds of thousands of them live in fairly isolated communities.*

Bajau People, Malaysia

While studying these people, researchers have discovered numerous physical differences between the Bajau and the indigenous individuals who live inland.

The Bajau subsistence lifestyle is based around diving for fish and shellfish. They do so while *holding their breath,* to depths of 200 feet or more.

Using a comparative, genomic study, a group of geneticists from the University of Copenhagen[348] showed that *natural selection on genetic variants in a particular gene increased the size of the spleen in the Bajau, providing them with a larger reservoir of oxygenated red blood cells.* An especially useful trait for deep diving humans.

The researchers also found evidence of *strong selection specific to the Bajau on another gene that affects the human diving reflex.*[CXXXIII] This variation also helps them to survive and reproduce in their environment.

With regard to the future of the human race, scientists are not agreed on how we will evolve. But the most likely scenario is *that*

[CXXXIII] The **diving reflex** is a clever physiological mechanism enabling the body to manage and tolerate a lower level of oxygen.

467

Darwinian evolution by natural selection *will not continue* for most Earth-bound humans.[CXXXIV]

The evolution of humankind will, however, continue. In fact, it will, *most likely, be much more rapid than in the past, and will be achieved by different, man-made means* rather than Natural Selection.

Genetic Engineering (GE) is one way that this new, rapid evolution will occur. Genetics is a swiftly evolving technology that holds great promise for the betterment of humankind.

It was inevitable that the detailed understanding of the human genome has led to the concept of Genetic Engineering. *This is the direct manipulation of a person's DNA to alter the organism's characteristics (phenotypes) in a particular way.*

In the near future, new gene editing tools, which will be much more precise than today's systems, will be used to make easy, rapid, safe genetic changes in progeny. As a result, *Darwin's concepts of random mutations and natural selection leading to new, better adapted species, will be obsolete.*

As discussed in the section *"Manipulating Intelligence: CRISPR-cas9A" on page 74,* the current, major tool for Genetic Engineering is called CRISPR-Cas9. *The letters stand for "Clusters of Regularly Interspaced, Short Palindromic Repeats" that are found in a DNA sequence. CRISPR was adapted from a naturally occurring, genome-editing system present in bacteria.* It was part of the bacteria's innate defense mechanism that they used to foil attacks on them by invading viruses or other foreign bodies.

Essentially, CRISPR is a specialized stretch of DNA that acts like a molecular homing device. When combined with Cas9 (CRISPR associated protein 9), a molecule that acts as a scissors, the two behave like a text editor in a word processing system.

[CXXXIV] **Darwinian evolution** may continue with humans *living on other planets, such as Mars,* because they will be a confined population living in dramatically different environmental conditions than on Earth. See the next section about exoplanetary colonies.

Together, the two molecules can repair, disable, or insert something new into a target gene. *This means researchers can easily alter DNA sequences and, thereby, change genes' functions.*

Today's technology, however, is not completely safe. CRISPR can confuse one gene for another, *and unintended results can occur.* Patients have accidently had the wrong gene selected resulting in the triggering of a deadly disease.

When this problem is resolved, most serious diseases in the future will be overcome by tweaking individual genomes or selecting specific embryos to avoid health problems. Lives will be improved and made to be significantly longer through Genetic Engineering.

Genetic Engineering, however, affects the Natural Selection process. People who *would have died and be taken out of the population* will now live to reproduce. This is yet another factor that thwarts Natural Selection.

Even more important, from an evolutionary standpoint, *is going to be the use of Genetic Engineering to create widespread, "designer babies" **and then the propagation of those genetic changes in the population through reproduction**.*

Much work is underway to improve gene editing technology, so mistakes are not made in gene selection. When this is accomplished in the relatively near future, the children of the very wealthy will be born with a base level of physical fitness, attractiveness, and intelligence that automatically places them in the top 1% of the human race.

There is much debate going on in the world about the moral correctness of the use of GE to alter the human genome and create designer babies. On the other hand, the bioethicists, Julian Savulescu and John Harris, have argued that it is not only a right but *a duty of all parents* to manipulate the genetic code for the improved future of their children. A concept they term "procreative beneficence."[349]

For all of these reasons, society will eventually push to have every child, not just those of the rich, genetically designed. ***And, as a result, over the next one hundred years, the human race will evolve by manufactured technologies more than it has for the last ten thousand years by Darwinian evolution.***

The Impact of the Singularity

The coming *Singularity*[CXXXV] will be even *more impactful on the evolution of the human race than Genetic Engineering. The emergence of the Superintelligence by mid-21ˢᵗ century will lead to explosive technological progress in many fields including genetics and machine intelligence itself.*

The results will be dramatic and unforeseen. Eventually the changes will reach heights far above what human beings can currently comprehend.

For example, *an entirely different mode of human evolution,* beyond Genetic Engineering, will be *computer-based processes that will be incorporated into our biological functioning. These changes will ultimately transcend our earthly bodies and we will become more machine-based and dramatically more long-lived.* **We will become Homo machina, the next species beyond Homo sapiens.**

Even more exciting is that much of this change will begin over the next fifty years, and then, according to the law of accelerating returns, speed up rapidly thereafter. *Therefore, it will occur within the lifetimes of many readers of this book.*

The major question: *can we control the rapid evolution that we are about to undergo, or will it run away from us with substantial, unintended consequences?*

The Brain/Machine/Cloud Interface

You may recall that in the movie, **The Matrix**, it was possible to upload detailed knowledge to a person's brain through a brain/machine interface attached to the back of the skull. This knowledge might be a language, the ability to fly a sophisticated airplane, or the skills of a martial arts expert. The subject could then use that ability in the virtual world of The Matrix.

Experts agree that such a scenario will likely become reality for mankind within the next fifty years. The advent of the Superintelligence with the Singularity, and the availability of the technology for a Brain/Machine/Cloud interface (B/M/CI), will be

---

[CXXXV] See Part III, **Artificial Intelligence,** "Waiting for the Singularity", page 323.

the means to a very significant cognitive enhancement for humans.

It will come about through the individual's direct access to a supercomputing processor in the "cloud"ᶜˣˣˣⱽᴵ that is operating artificial intelligence algorithms associated with a massive data base of knowledge.

There are multiple approaches to achieving a B/M/CI including a probe inserted directly into the brain as seen in The Matrix. In fact, *brain computer interface technology has existed for several decades.* It tends to be used for some medical applications. *So there is experience with the approach in the medical community.*

John Donoghue, a Brown University professor of neuroscience and engineering, developed such a medical connection in 2000 and continues his work in this field.[350] According to Donoghue, there are more than 300,000 people who already have some form of neural interface. *One example is a deep brain stimulator used to treat Parkinson disease.*

A *major barrier* to making a more sophisticated B/M/CI is that *we do not have a perfect understanding of how the brain works.* Neuroscience is well-developed but, according to *Christof Koch, chief scientist at the Allen Institute for Brain Science:*

> *"The brain is the most complex piece of highly organized active matter in the known universe. We do not understand how large-scale neural activity is organized to give rise to thoughts, percepts, consciousness and actions."*

ᶜˣˣˣⱽᴵ The term **"cloud"** refers to *cloud computing.* This is the delivery, over the Internet (cloud), of computing services, including servers, storage, databases, networking, software, analytics, etc.,

Dr. Christof Koch

Elon Musk, the well-known entrepreneur, has a company, *Neuralink*, that has been developing and testing a Brain/Machine Interface with the goal of ultimately connecting humans to the Superintelligence. *Neuralink is the closest technology to working in that application and will likely be the first to succeed at some level.*

A founding member of Neuralink, Philip Sabes, said:

> "If it were a prerequisite to understand the brain in order to interact with the brain in a substantive way, we'd have trouble. **But it's possible to decode all of those things in the brain without truly understanding the dynamics of the computation in the brain**.

> "Being able to read it out is an engineering problem. Being able to understand its origin and the organization of the neurons in fine detail in a way that would satisfy a neuroscientist to the core—that's a separate problem. **And we don't need to solve all of those scientific problems in order to make progress.**"

Christof Koch who understands the limitations of neuroscience, has said of Neuralink:

> "I admire them for their long-term vision, drive and ambition. It will take time though."

The goal of Neuralink is to implant a computer chip, roughly the size of a large coin, into a human skull, and then connect it with microfine wires to the brain. It will be done by using a robot surgeon to make the procedure as precise as possible. The chip, which Neuralink calls the "Link," will wirelessly connect the brain to the digital world, starting by connecting it to a "smart phone".

Musk is spearheading this approach for a good reason. He has long been warning of the potential dangers when the Superintelligence appears. He has said:[351]

> *"A.I. does not need to hate us to destroy us. We would roll over an anthill that's in the way of a road. You don't hate ants. You're just building a road.*

> *"I created [Neuralink] specifically to address the A.I. symbiosis problem, which I think is an existential threat."*

So far, the Neuralink test subjects have been animals. One drawback to the brain implant approach for a Brain/Machine/Cloud interface (B/M/CI) it that it is rather frightening to most people, i. e. having a robot surgeon cut a hole in your skull and insert a neural microchip, sensor array with attached wire probes into your brain. Moreover, *the number of probes is highly limited compared to the eighty-six billion neurons in a human brain.*

On the other hand, the ultimate route for achieving the B/M/CI may well be *infinitesimally small, medical nanorobots* that will be injected directly into your bloodstream, with no more invasiveness than a vaccine injection. **These miniscule devices are called neural nanorobots (or nanobots[CXXXVII]).** Other groups of scientists are pursuing this approach.

Neural nanorobotics will provide a non-destructive, real-time, secure, long-term, and virtually autonomous, *in vivo system* that can realize the first fully functional, human Brain/Machine/Cloud interface.[352] *There are many hurdles to pass, but scientists are confident that the barriers will be overcome within the next forty years.*

[CXXXVII] **Nanobot**, is a popular term for *molecules* designed with a unique property that enables them to be programmed to conduct a specific task.

The Futurist, Ray Kurzweil, was the first person to suggest neural nanobots as the means to connect a brain to the cloud.[353] Since then, much work has been done on developing this technology, and it appears to be entirely feasible.

Recently Kurzweil said,

> "*We'll have nanobots that... connect our neocortex to a synthetic neocortex in the cloud... Our thinking will be a.... biological and non-biological hybrid.*"

An *international group of twelve researchers* came together recently[354] and predicted that, before mid-21st century, neural-nanorobots will provide a secure, real-time Brain/Machine/Cloud interface (B/M/CI). Such B/M/CI systems will dramatically alter human/machine communications and create significant human cognitive enhancement. *And the key to achieving this reality lies with the development of nanotechnology.*

Richard Feynman was a renowned physicist from the California Institute of Technology. In 1959, he gave a talk that described the possibility of synthesizing objects by the direct manipulation of atoms or molecules.[355] **This prospect became known as nanotechnology.**

"Nano" refers to the scale of the devices. That is, in the range of a nanometer, or 10^{-9} meters, or one billionth of a meter. A nanometer *is just twenty times wider than the diameter of a hydrogen atom.*

As another indication of the scale we are talking about, imagine that you were shrunk to be one nanometer tall. Then, stand next to a man of average height. Relative to you, his height would be *three times higher than the distance from the Earth to the Moon!*

It took until the 1980s before technology advanced far enough for researchers to actually be able to work at the nanometer level. A breakthrough was made in 1981 with the invention of the *scanning-tunneling microscope*[CXXXVIII] *(STM).* This device allowed

---

[CXXXVIII] **Scanning tunneling microscope** - A microscope that scans the surface of a sample with a beam of electrons. A narrow channel of "tunneling" electrons flows between the sample and the beam. The

scientists to see individual atoms and bonds and even manipulate them. The STM became the first step in being able to design and build *nanobots for medical applications.* Today's best STMs have a resolution of 0.05 nanometers, the diameter of a hydrogen atom!

Professor Richard Feynman

Initially, there were some concerns about the toxicity of some the nanomaterials being considered for neural nanobots.[356] This has, however, been resolved with regard to the development of the neural nanobot.

Advanced medical nanobots will likely be fabricated using *diamondoid* materials. Diamonoids, or nanodiamonds as they are sometimes called, occur naturally in petroleum deposits and have been extracted and purified into large, pure crystals. So the material is generally available.

Diamondoid is non-toxic, and provides the greatest strength, durability, and reliability for use in the *in vivo environment.* It features extremely smooth surfaces, lessening the possibilities of triggering the body's immune system.

A group of international scientists have provided a detailed, scientific prediction about the future of neural nanorobotics:[357]

---

result is a three-dimensional image of atoms in the sample, showing their topography and structure.

"Neural nanorobotics may also enable a Brain/Machine/Cloud interface with controlled connectivity between neural activity and external data storage and processing, via the direct monitoring of the brain's 86 billion neurons and 200 trillion synapses.

"Subsequent to navigating the human vasculature, three species of neural nanorobots (endo-neurobots, gliabots, and synaptobots) could traverse the blood–brain barrier (BBB), enter the brain parenchyma (the functional tissue of the brain), *ingress into individual human brain cells, and autoposition themselves at the axon initial segments of neurons (endo-neurobots), within glial cells (gliabots), and in intimate proximity to synapses (synaptobots).*

"They would then wirelessly transmit up to ~6 × 10^{16} bits per second of synaptically processed and encoded human–brain electrical information via auxiliary nanorobotic fiber optics (30 cm^3) with the capacity to handle up to 10^{18} bits/sec and provide rapid data transfer to a cloud-based supercomputer for real-time, brain-state monitoring and data extraction.

"A neural-nanorobotically enabled human Brain/Cloud Interface might serve as a personalized conduit, allowing persons to obtain direct, instantaneous access to virtually any facet of cumulative human knowledge.

"Other anticipated applications include myriad opportunities to improve education, intelligence, entertainment, traveling, and other interactive experiences."

To have B/MCIs widely available using neural nanobots, it will be necessary to manufacture them by the trillions at low cost. This problem has been considered by scientists and they believe that it will be possible in the future with molecular manufacturing processes.[358]

Molecular manufacturing provides the capability to build complex structures at the atomic level. But the technology will require significant advances in nanotechnology before it can be carried out.

Eventually, a nanofactory will produce highly advanced products at low costs and in enormous quantities. At some point, the nanofactories will gain the ability to produce other nanofactories. Then production will only be limited by considerations such as the required input materials, energy, and software.

An ongoing international "Nanofactory Collaboration" headed by Robert Freitas and Ralph Merkle, two of the pioneers in nanotechnology, has the primary goal of constructing the world's first nanofactory. *The collaborative is coordinating a joint experimental and theoretical* effort involving direct cooperation among dozens of researchers at *multiple institutions in four countries.* This will allow *the mass manufacture of advanced autonomous diamondoid neural nanorobots for both medical and non-medical applications.*

The neural nanorobotically mediated, Brain/Cloud Interface systems will enable significantly increased human intelligence. This will be done with the hybrid, biological/non-biological systems, using new forms of AI, which will considerably expand human memory capabilities, and cognition. Some might say that *we will become "cyborgs".* [CXXXIX]

Popular culture is filled with the idea of cyborgs. The 'seventies television series, **The Six Million Dollar Man** and **The Bionic Woman** were early examples. Then came Darth Vader from the **Star Wars** movies, and **Robocop.**

[CXXXIX] The term **'cyborg'**—derived from *cybernetic organism*—was coined in 1960 by Manfred Clynes and Nathan Kline. In the popular image, the *cyborg stands for the merging of the human and the machine.*

Robocop

The idea of a cyborg has come to mean all types of human-machine combinations and all sorts of technological and biotechnological enhancements or modifications of the human body.

These fictional characters had replaced many body parts with enhanced "smart" replacements. An actual example of current technology is the Ottobok C-Leg.

Sensors placed in the artificial C-Leg, combined with a microprocessor-controlled knee, aids the wearer in walking with a natural gait. *The product is considered by some to be the first step towards the next generation of real-world, cyborg applications.*[359]

It is highly likely that over the next one hundred years, most of humankind will evolve into super intelligent, biomechanically advanced cyborgs that will live productively for well over one hundred years, or even much longer as the technology progresses. We will have changed so much through genetic engineering and our symbiosis with the superintelligence that we will be a new species, Homo machina.

A Final Word

Given that planets similar to Earth are plentiful all over the galaxy, scientists have assumed that the Cosmos is teaming with life. Yet, the Fermi Paradox holds true, *"Where is everybody?"*.

As far as we know from all of our sophisticated observations, *there is no consciousness or intelligence in all of Space-Time except for us.* Yet, there should be.

The history of life on Earth has certainly proved that there is a *biological imperative for life to grow and spread as far as possible.* This imperative continues with humankind. And it should have been true for any lifeforms that developed on other planets.

There should have been numerous civilizations throughout the cosmos that developed over the fourteen billion years since the Big Bang. And they should have made their marks throughout the galaxy in ways that we can see. Yet there is no evidence that they are there. They were stopped by some "Great Filter".

Now, we are on the verge of leaving Earth and settling on Mars, another planet in our solar system. *Could humankind be the answer to the Fermi Paradox?* We may be the first civilization in the galaxy to get past the Great Filter and to reach this point, where we can move out and populate the galaxy and then the Universe over the next millions and billions of years.

Our intelligence, natural and artificial, is growing rapidly and has resulted in amazing technological progress over the last one hundred years. And this progress will continue to grow exponentially at an even faster rate. Soon, we will become part of the Superintelligence.

As the only known conscious, intelligent creatures in the Universe, we have a fundamental obligation to spread these attributes throughout the Cosmos. *And with our growing technological capabilities, we will.*

The End

Index

Supernova, 369
superposition, 126, 127, 128, 132, 133, 266, 267
survival of the fittest, 158
Suzana Herculano-Houzel, 19, 90, 167, 168, 194, 203
Swarm Intelligence, 307
Swarm Robotics, 308
synapse, 92, 93, 95, 96, 97, 98
Synapses, 92, 98
synaptic cleft, 95, 96
Synaptic pruning, 97, 98
synaptogenesis., 97
temporal lobe, 83, 86, 99
Temporal lobe, 83
terraform, 501
Tesla automobiles, 309
The Anthropic Cosmological Principle, 144
The Attention Schema Theory, 135
The Bell Curve, 35, 53, 54, 55, 56, 57, 59, 68
The Imitation Game, 245, 248
The Integrated Information Theory, 137
the interneuronal communication scheme, 93
The Singularity, 22, 239, 341, 454
The Stanford-Binet Intelligence Scales, 32
the theory of mind, 137, 337, 338, 339, 340, 341
The Woodcock-Johnson Tests of Cognitive Abilities, 32

Theory of Evolution by Natural Selection, 153
Theory of Mind, 238, 338
Theory of Relativity, 36, 128, 268
tractograms, 91
traits, 21, 35, 40, 41, 42, 43, 45, 66, 67, 103, 153, 160
trilobite, 156
trucking, 313
Turing machines, 244
Turing Test, 240, 245, 248, 334
Turing's *"stored-program concept"*,, 243
TuSimple, 313, 314
twin study, 67, 69
UFO, 20, 21, 400, 403, 404, 411, 412, 414, 415, 416, 417, 418, 419, 420, 421, 428, 431, 437, 438, 439
ultraviolet catastrophe, 129
University of Cambridge, 230, 245
University of St. Andrews, 109, 435
use of tools, 172, 177, 191, 460
visuospatial, 86
waggle dance, 221, 222
Washoe, 178, 179, 180, 181
Watson and Crick, 64
wave function collapse, 126
Wechsler Adult Intelligence Scale, 32, 33, 88
William King, 464
Williams-Kilburn, 248
Wilson Effect, 70
wolves, 160, 203, 204
Wrens, 246

End Notes

1. "Mainstream science on intelligence: an editorial with 52 signatories, history, and bibliography", Gottfredson, LS, **Intelligence,** 24: 13–23 (1997).

2 "The Measurement of Intelligence: An Explanation of and a Complete Guide for the Use of the Stanford Revision and Extension of the Binet-Simon Intelligence Scale", Terman, L., **Houghton Mifflin**, (1916).

3 "Individually Administered Intelligence Tests", Machek, G., **Human Intelligence**, (Summer 2006), https://www.intelltheory.com/intelligenceTests.shtml

4 **The Abilities of Man: Their Nature and Measurement**, Spearman, Charles, **Macmillan,** (1927).

5 "Best practices in exploratory factor analysis", Costello, A. B., Osborne, J., **Practical Assessment, Research and Evaluation**, Volume 10, Article 7, (2005).

6 "Theory of fluid and crystallized intelligence: A critical experiment", Cattell R.B., **Journal of Educational Psychology**, 54: 1-22, (1963), DOI: 10.1037/h0046743.

7 "IQ Scale Meaning", van Thiel, E., **123 Test,** (May 21, 2019) https://www.123test.com/interpretation-of-an-iq-score.

8 "Intelligence (IQ) as a Predictor of Life Success", Firkowska-Mankiewicz, A., et al, **International Journal of Sociology,** pp. 25-43, (2002)

9 "High IQ in Early Adolescence and Career Success in Adulthood: Findings from a Swedish Longitudinal Study", Bergman, **Research in Human Development**, Volume 11, 2014 – Issue 3: Pages 165-185, Published online: (15 Aug 2014)

10 "The association between IQ in adolescence and a range of health outcomes at 40 in the 1979 US National Longitudinal Study of Youth", Der, G., et al, **Intelligence**, 37(6): 573–580, (November 2009)

11 "Grit: Perseverance and passion for long-term goals," Duckworth, A., et al, **Journal of Personality and Social Psychology**, Vol 92(6), 1087-1101, (Jun 2007)

12 "The Eighty Five Percent Rule for optimal learning", Wilson, R.C., *et al,* **National Communications,** 10, 4646 (2019). https://doi.org/10.1038/s41467-019-12552-4

[13] "Role of Optimism on Employee Performance and Job Satisfaction", Mishra, Uma, et al, **Prabandhan: Indian Journal of Management,** 9(6): 35, (June 2016).

[14] **Emotional Intelligence: Why it can matter more than IQ,** Goleman, Daniel, **Bantam Books**, (1995).

[15] **Hereditary Genius**, Galton, Sir Francis, (1869), http://galton.org/books/hereditary-genius/text/pdf/galton-1869-genius-v3.pdf

[16]. "The History of Twins, As a Criterion of the Relative Powers of Nature and Nurture", Galton, Francis, **Fraser's Magazine, 12,** 566-576 (1876)

http://galton.org/essays/1870-1879/galton-1875-history-twins.pdf

[17] **Inquiries into Human Faculty and its Development**, Galton, Sir Francis, (1883). http://galton.org/books/human-faculty/text/human-faculty.pdf

[18] "The Eugenic Origins of IQ Testing: implications for Post-Atkins Litigation", Reddy, Ajitha, **DePaul Law Review,** Volume 57, Issue 3, Symposium-Protecting a National Moral Consensus: Challenges in the Application of Atkins v. Virginia, (Spring 2008)

[19] "How Much Can We Boost IQ and Scholastic Achievement", Jensen, Arthur, **Harvard Educational Review**: Vol. 39, No. 1, pp. 1-123, (April 1969). https://doi.org/10.17763/haer.39.1.l3u15956627424k7

[13] **The g Factor: The Science of Mental Ability**, Jensen, Arthur R., **Greenwood Publishing,** Westport, Connecticut, (1998). ISBN 0-2759-6103-6. 88.

[21] "Arthur Jensen dies at 89; his views on race and IQ created a furor", Woo, Elaine, **Los Angeles Times**, November 2, 2012
[22] Ibid

[24] **The Bell Curve: Intelligence and Class Structure in American Life,** Richard J. Herrnstein and Charles Murray, **Free Press**, 1994
[25] "The stunning dishonesty of Charles Murray", Joan Walsh, **Salon**, March 18, 2014
[26] "Curveball", Gould, Stephen Jay, **The New Yorker**, (November 28, 1994)

[27] "Intelligence: Knowns and Unknowns", U. Neisser, et al, **American Psychologist**, 51(2), 77–101, (1996).

https://doi.org/10.1037/0003-066X.51.2.77

[28] "'The Bell Curve' 20 years later: A Q&A with Charles Murray", Goodnow, Natalie, **The American Enterprise Institute**, Blogpost, October 2014

[29] "The cognitive impact of the education revolution: A possible cause of the Flynn Effect on population IQ", Baker, David P., et al, **Intelligence,** 49: 144-158, (March 2015).

[30] "Requiem for nutrition as the cause of IQ gains: Raven's gains in Britain 1938–2008", Flynn, James R., **Economics and Human Biology**, 7 (1): 18–2, (March 2009).

[31] "The generational intelligence gains are caused by decreasing variance in the lower half of the distribution: Supporting evidence for the nutrition hypothesis", Colom, R., et al, **Intelligence** 33 (1): 83–91. (2005). doi:10.1016/j.intell.2004.07.010

[32] "Rising Scores on Intelligence Tests", Neisser, Ulric, **American Scientist**, 85 (5): 440–47. (1997).

[33] **What Is Intelligence: Beyond the Flynn Effect**, Flynn, James R., **Cambridge University Press,** (2009). ISBN 978-0-521-74147-7

[34] "The generational intelligence gains are caused by decreasing variance in the lower half of the distribution: Supporting evidence for the nutrition hypothesis", Colom, R, et al, **Intelligence,** 33 (1): 83–91, (2005).

[35] "Parasite prevalence and the worldwide distribution of cognitive ability", Eppig C, et al, **Proceedings of Biological Science**, 277 (1701):3801-3808, (December2010)

doi:10.1098/rspb.2010.0973. PMC 2992705. PMID 20591860

[36] "Rising Scores on Intelligence Tests", Neisser, Ulric, **American Scientist,** 85 (5): 440–47, (1997).

[37] "The end of the Flynn effect?": A study of secular trends in mean intelligence test scores of Norwegian conscripts during half a century", Sundet, J., et al, **Intelligence,** 32 (4): 349–62, (2004). doi:10.1016/j.intell.2004.06.004

[38] "Secular declines in cognitive test scores: A reversal of the Flynn Effect", Teasdale, TW, Owen, DR, **Intelligence,** 36(2): 121–26, (2008). doi:10.1016/j.intell.2007.01.007.

[39] "British teenagers have lower IQs than their counterparts did 30 years ago", Gray, Richard, **The Telegraph,** London, England, (February 7, 2009).

[40]. "The decline of the world's IQ", Lynn, Richard. Harvey, John, **Intelligence,** 36 (2): 112, (March 2008).

doi:10.1016/j.intell.2007.03.004. ISSN 0160-2896.

[41] "The History of Twins, As a Criterion of the Relative Powers of Nature and Nurture", Galton, Francis, **Fraser's Magazine**, 12 566–76. (1875): http://galton.org/essays/1870-1879/galton-1875-history-twins.pdf.

[42] "Top 10 Replicated Findings from Behavioral Genetics", Plomin, Robert, et al, **Perspectives in Psychological Science**, 11(1): 3–23d, (Jan 2016). doi: 10.1177/1745691615617439

[43] "Behavioral genetics of cognitive ability: A life-span perspective", McGue M, Bouchard TJ, Jr., Iacono WG, Lykken DT., Plomin R, McClearn GE, editors, **Nature, Nurture, and Psychology. American Psychological Association**, Washington, DC: pp. 59–76, (1993).

[44] "Genome-wide association meta-analysis of 78,308 individuals identifies new loci and genes influencing human intelligence", Sniekers, S., Stringer, S., Watanabe, K. et al, **Nature Genetics**, 49, 1107–1112 (2017). https://doi.org/10.1038/ng.3869

[45] "Meta-analysis: potentials and promise", Egger, M, et al, BMJ, 315:1371–4, (1997).

[46] Ibid

[47] "The new genetics of intelligence", Plomin, R., von Stumm, S., **Nature Reviews Genetics**, 19(3): 148–159. (2018 Mar) doi: 10.1038/nrg.2017.104

[48]. "Genome-wide association meta-analysis in 269,867 individuals identifies new genetic and functional links to intelligence," Savage J. E., Jansen P. R., Stringer S., Watanabe K., Bryois J., de Leeuw C. A., et al., **Nature Genetics**, 50, 912–919. (2018). 10.1038/s41588-018-0152-6
[49] Ibid

[50] "The new genetics of intelligence", Plomin, R. and von Stumm, S., **Nature Reviews Genetics**, 19(3): 148–159. (March 2018), doi: 10.1038/nrg.2017.104

[51] Ibid

[52] "DNA tests for IQ are coming, but it might not be smart to take one", Regala, Antonio, **MIT Technology Review**, (April 2, 2018)

[53] "CRISPR base editors can induce wide-ranging off target edits in RNA", Joung, J.K., Mass General Hospital Journal, (May 28, 2020)

[54] "In vivo brain size and intelligence", Willerman, L, et al, **American Psychological Association, Intelligence**, 15(2), 223–228, (1991).

[55] "Big-brained people are smarter: A meta-analysis of the relationship between in vivo brain volume and intelligence", McDaniel, M.A., **Intelligence** (33) 337 – 346, (2005).

[56] **The Human Advantage**, Suzana Herculano-Houzel, Vanderbilt University, **The MIT Press**, Cambridge, Massachusetts, (2016)

[57] "Synaptic pruning in development: a computational account", Chechik, G, et al, **Neural Computation,** 10(7), 1759–1777, (1998)

[58] "The synaptic pruning hypothesis of schizophrenia promises and challenges", Keshavan, M., et al, **World Psychiatry,** 19 (1): 110–111, (2020).

[59] "Large and fast human pyramidal neurons associate with intelligence", Goriounova, Natalia A., et al, Vrije Universiteit Amsterdam, The Netherlands, **eLife,** 7:e41714, (December 18, 2018). DOI: 10.7554/eLife.41714,

[60] "Neural coding of natural stimuli: information at sub-millisecond resolution", Nemenman, I., et al, **PLoS Computational Biology,** 4:e1000025, (Mar 7, 2008). https://doi.org/10.1371/journal.pcbi.1000025

[61] "The Parieto-Frontal Integration Theory (P-FIT) of intelligence: Converging neuroimaging evidence", Jung, Rex E., et al, **Behavioural And Brain Sciences** 30, 135–187, (2007).

[62] Ibid

[63] "The neuroscience of human intelligence differences", Deary, Ian J., et al, **Nature Reviews, Nuroscience**, volume 11, 201-211, (March 2010).

[64] "Herding in humans", Ramsey M.R., et al, **Trends in Cognitive Sciences,** Volume 13, Issue 10, Pages 420-428, (October 2009).

[65] "What smart bees can teach humans about collective intelligence", Toyokawa, Wataru, School of Biology, University of St Andrews, **The Conversation**, January 29, 2019.

[66] "The Neural Correlates of Consciousness", Frackowiak, R, et al, pgs. 269-301, **Human Brain Function,** Elsevier, (2004)

[67] "Near-death experience in survivors of cardiac arrest: a prospective study in the Netherlands", Dr Pirn van Lommel, MD, et al, **Lancet**, Volume 358, Issue 9298, P2039-2045, (December 15, 2001).

[68] **Consciousness Beyond Life: The Science of the Near-Death Experience**, van Lommel, Pim, **Harper One**, (June 8, 2010).

[69] "The hard problem of consciousness", David J. Chalmers, **The Blackwell Companion to Consciousness.** Blackwell (2007), Max Velmans & Susan Schneider (eds.)

[70] **Philosophiæ Naturalis Principia Mathematica**, Isaac Newton, Rough Draft Printing, (2011)

[71] "Consciousness in the universe: A review of the 'Orch OR' theory'", Stuart Hameroff, Roger Penrose, **Physics of Life Reviews**, Volume 11, Issue 1, Pages 39-78, (March 2014)

[72] "The importance of quantum decoherence in brain processes", Max Tegmark, **Physical Review E**, 61, 4194-4206, (October 25, 1999)

[73] "The attention schema theory: a mechanistic account of subjective awareness", Taylor W. Webb and Michael S.A. Graziano, Princeton Neuroscience Institute, **Frontiers in Psychology**, Vol. 6, Article 500, (23 April 2015).

[74] "Integrated information theory: from consciousness to its physical substrate", Tononi, G., et al, **Nature Reviews: Neuroscience**, volume 17, pages 450–461 (2016)

[75] "Conscious realism and the mind-body problem", Donald D. Hoffman, **Mind and Matter,** 6: 87-121, (2008)

[76] **The Anthropic Cosmological Principle**, John D. Barrow, et al, **Oxford University Press**, (1985)

[77] **The Hidden Spring: A Journey to the Source of Consciousness**, Mark Solms, **W.W. Norton & Co**. New York, New York, (2021)

[78] **The Origin of Species**, Charles Darwin, *first published 1859*, **Enhanced Media Publishing**, Los Angeles, California, (2017)

[79] "Evolutionism(s) and Creationism(s)", Brosseau, Olivier; Silberstein, Marc, Handbook of Evolutionary Thinking in the Sciences, Dordrecht: Springer. pp. 881–96, 2015

[80] "The evolution of George Gilder", Joseph P. Kahn, Staff, **The Boston Globe**, (July 27, 2005)

[81] **Darwin's Doubt**, Stephen C. Meyer, *Harper One*, (2013)

[82] "Peer Reviewed Articles supporting Intelligent Design", **Discovery Institute, Center for Science and Culture**, https://www.discovery.org/id/peer-review.

[83] "Evolution of cognition", Jayden van Horik, Nathan J. Emery, **Wires Cognitive Science**, (11 April 2011) https://doi.org/10.1002/wcs.144

[84] "What Is Animal Intelligence?", Jonathan Carey, **Sentient Media**, (September 26, 2018).

[85] "Evolution in the Social Brain", Dunbar, R. I., Shultz, Susanne, **Science**, Volume 317, Issue 5843, pp. 1344, (September 2007)

[86] "Cognitive adaptations of social bonding in birds", Emery, Nathan J., et al, **Philosophical Transactions of the Royal Society of London B: Biological Sciences**. 362 (1480): 489–505, (2007-04-29)

[87] "Memory for food caches: not just for retrieval", Male, Lucinda H.; Smulders, Tom V., **Behavioral Ecology**, 18 (2): 456–459, (2007-03-01).

[88] "Evolutionary Biology of Animal Cognition", Dukas, Reuven, **Annual Review of Ecology, Evolution, and Systematics,** 35: 347–374, (2004)

[89] "Insightful problem solving and creative tool modification by captive non-tool-using rooks", Bird, Christopher D.; Emery, Nathan J, **Proceedings of the National Academy of Sciences**, 106 (25): 10370–10375, (2009-06-23).

[90] "Evolution of cognition", van Horik, Jayden, Emery, Nathan J., Wiley Interdisciplinary Reviews: **Cognitive Science**. 2 (6): 621–633, (2011-11-01).

[91] "Monkeys Washing Potatoes", K, Alfred, **alfre.dk**, (2018-03-26).

[92] Ibid

[93] "Cultures in chimpanzees", J. Goodall, et al, **Nature**, 399(6737):682-5, (June 17, 1999).

[94] **The Human Advantage**, Suzana Herculano-Houzel**, The MIT Press**, Cambridge, Massachusetts, 2016.

[95] **The Abilities of Man: Their Nature and Measurement**, Spearman, C., New York: Macmillan, (1927).

[96] "Best practices in exploratory factor analysis", Anna B. Costello Jason Osborne, , **Practical Assessment, Research and Evaluation**, Volume 10, Article 7, (2005).

[97] "General intelligence in another primate: individual differences across cognitive task performance in a New World monkey", Banerjee, K., et al, **Public Library of Science One**, 4(6), e5883, (2009).

[98] "A general intelligence factor in dogs", Rosalind Arden, Mark J. Adams, **Intelligence**, 55 (March–April 2016) 79–85

[99] "The evolution of primate general and cultural intelligence", Reader, S. M., Hager, Y., Laland, K. N., **Philosophical Transactions of the Royal Society B: Biological Sciences**, 366(1567), 1017-1027, (2011).

[100] "How general is cognitive ability in non-human animals? A meta-analytical and multi-level reanalysis approach", Poirier, Marc-Antoine, et al, **Proceedings of the Royal Society B: Biological Sciences**, 287 (1940): 20201853, (2020-12-09).

[101] "Inferences About the Location of Food in the Great Apes (Pan paniscus, Pan troglodytes, Gorilla gorilla, and Pongo pygmaeus)", Call, J., et al, **Journal of Comparative Psychology**, 118(2), 232–241, (2004).

[102] "Orangutans play video games (for research) at Georgia zoo", Turner, Dorie, **USA Today**, 12 April 2007.

[103] "Cooperative problem solving by orangutans (*Pongo pygmaeus*)", Chalmeau, Raphaël, et al, **International Journal of Primatology,** 18 (1): 23–32, (1997).

[104] "Calculated reciprocity after all: computation behind token transfers in orangutans". Dufour, V. et al, **Biology Letters,** 5 (2): 172–75, (23 December 2008) doi:10.1098/rsbl.2008.0644.

[105] **A morally deep world: an essay on moral significance and environmental ethics,** Johnson, Lawrence E., **Cambridge University Press**, p. 27, ISBN 978-0-521-44706-5, (1993)

[106] **Drawing the line: science and the case for animal rights**, Wise, Stephen M., **Basic Books**, p. 200, ISBN 978-0-7382-0810-7, (2003)

[107] **Nim Chimpsky, the chimp who would be human,** Elizabeth Hess, **Bantam Dell**, New York, New York, (2008)

[108] "Vocabulary size and auditory word recognition in preschool children", Law, Franzo, **Applied Psycholinguistics,** 38(1): 89–, (January 2017)

[109] "Scientific Publications". **The Gorilla Foundation.** 52 articles, 13 books. Retrieved on (10/31/2021).

[110] "What it's like to be interviewed for a job by Koko the gorilla: 'She had a lot to say'", Jensvold, Mary Lee, **Science (Science Now), Los Angeles Times (Interview),** Interviewed by Deborah Netburn, (June 22, 2018).

[111] "The Hidden World of Whales", Craig Welch, **National Geographic**, (May, 2021)

[112]. "The social and cultural roots of whale and dolphin brains", Kieran, C. R., Fox, Michael, Shultz, Susanne, et al, **Nature Ecology & Evolution,** (2017).

[113]. "Relative brain sizes and cortical surface areas of odontocetes", Ridgway, S.H. and Brownson R.H, **Acta Zoologica Fennica,** 172, 149–152, (1984).

[114] "Total neocortical cell number in the mysticete brain", Eriksen, N. and Pakkenberg, B., **Anatomical Record,** 290, 83–95, (2007)

[115] "Smarter than the average chimp", Dingfelder, Sadie F., **American Psychological Association**, , Vol 35, No. 9, Print version: page 72, (October 2004).

[116] "Why Dolphins Are the Most Intelligent Ocean Creature", Islam, N., **Awesome Ocean**, The Ultimate Guide: Dolphins, Discovery Communications Inc., (1999)..

[117] "How smart are killer whales? Orcas have 2nd-biggest brains of all marine mammals", Spear K., **Phys.org,** (March 8 2010).

[118] "Killer whales communicate in distinct 'dialects'", Dayton, L., **New Scientist,** (10 March 1990).

[119] "Differences in acoustic features of vocalizations produced by killer whales cross-socialized with bottlenose dolphins", Musser, W.B., et al, **Journal Acoustic Society of America**, 136(4):1990-2002, (2014). doi:10.1121/1.4893906

[120] "Mirror image processing in three marine mammal species: Killer whales (Orcinus orca), false killer whales (Pseudorca crassidens) and California sea lions (Zalophus californianus)", Delfour F, Marten K., **Behavioral Processes,** 53(3):181-190, (2001).

[121]"Cognitive behaviour in Asian elephants: use and modification of branches for fly switching", Hart, B., et al, **Animal Behaviour**, 62:839-847, (2001).

[122] "Elephants can determine ethnicity, gender, and age from acoustic cues in human voices", McComb, Karen, et al, **Proceedings of the National Academy of Sciences**, (March 2014).

[123] "Asian elephants (Elephas maximus) reassure others in distress", Joshua M. Plotnik, et al, **Peer J, Life & Environment**,2:e278 (February 18,2014). https://doi.org/10.7717/peerj.278

[124] "Research shows how truly empathic elephants can be", **www.thespiritscience.net**, (August 12, 2016)

[125] "Birds have primate-like numbers of neurons in the forebrain", Seweryn Olkowicz, Suzana Herculano-Houzel, et al, **Proceedings of the National Academy of Sciences**, 113 (26) 7255-7260, (June 28, 2016).

[126] "This Video of a Wild Crow Using the Urban Landscape to Its Advantage is Incredible", Sir David Attenborough, (August 26, 2020).

https://twistedsifter.com/videos/wild-crow-uses-urban-landscape-to-its-advantage/

[127] "Complex cognition and behavioural innovation in New Caledonian crows", Alex H Taylor et al, **Proceedings of the Royal Society B: Biological Sciences**, Volume 277, Issue 1694, Pgs. 2637-2643, (Sept. 7, 2010).

[128] "Brain imaging reveals neuronal circuitry underlying the crow's perception of human faces", John M. Marzluff, et al, **Proceedings of the National Academy of Science**, 109 (39) 15912-15917 September 25, 2012.

[129] "Sensory and Working Memory Representations of Small and Large Numerosities in the Crow Endbrain", Ditz, Helen, et al, **The Journal of Neuroscience** 36(47):12044-12052, (November 2016).

[130] "Grey parrots use inferential reasoning based on acoustic cues alone", Christian Schloegl, et al, **Proceedings of the Royal Society of London**, Series B, 479, 4135-4142, (08 August 2012).

[131] "**Alex & Me**: *How a Scientist and a Parrot Discovered a Hidden World of Animal Intelligence--and Formed a Deep Bond in the Process*", Irene Pepperberg, **Harper Collins Publishers**, New York, N.Y., (2008).

[132] **Drawing the Line**, Wise, Steven M., Cambridge, Massachusetts: **Perseus Books**, p. 106. ISBN 0-7382-0340-8, (2002).

[133] "An Interview with Alex, the African Grey Parrot", Wong, Kate, **Scientific American,** (12 September 2007).

[134] "Talking with Alex: Logic and speech in parrots", Irene Pepperberg, **Scientific American,** pg. 61-65, (1998)

[135] "Farewell to a famous parrot", Chandler, David, **Nature** (11 September 2007). https://doi.org/10.1038/news070910-4.

[136] "The effects of domestication and ontogeny on cognition in dogs and wolves", Michelle Lampe, et al, **Scientific Reports**, volume 7, Article number: 11690, (2017).

[137] "Remembering Chaser, the 'Smartest Dog in the World'", Abbie Mood, **American Kennel Club**, (August 5, 2019)

[138] "Word Learning in a Domestic Dog: Evidence for "Fast Mapping"", Kaminski, Juliane, et al, **Science** (New York, N.Y.)., 304,1682-3, (2004).

[139] "The efficacy of the model–rival method when compared with operant conditioning for training domestic dogs to perform a retrieval–selection task", Sue McKinley, Robert Young, **Applied Animal Behaviour Science,** Volume 81, Issue 4, Pgs. 357-365, (21 May 2003).

[140] **The Intelligence of Dogs**, Coren, Stanley, Free Press, New York, N.Y., (1996).

[141] "A general intelligence factor in dogs", Rosalind Arden, Mark J. Adams, **Intelligence**, 55 (March–April 2016) 79–85

[142] "A general intelligence factor in dogs", Arden, R, Adams, M., **Intelligence**, Elsevier, vol. 55, pp. 79-85, (2016)

[143] **Octopus: The Ocean's Intelligent Invertebrate**, Anderson, Roland; Mather, Jennifer A., Timber Press, 2010.

[144] "Octopuses found to learn from one another," Dr. Graziano Florito, Dohrn Zoological Institute, Naples, Italy, **New York Times**, Sec. A page 16, April 24, 1992.

[145] **Octopus: The Ocean's Intelligent Invertebrate**, Anderson, Roland; Mather, Jennifer A., Timber Press, 2010.

[146] "A draft genome sequence of the elusive giant squid, Architeuthis dux", Rute R da Fonseca, **GigaScience**, Volume 9, Issue 1, January 2020.

[147] "Consensus decision-making in animals", Conradt, L. and Roper, T.J., Trends in Ecology and Evolution, 20, 449–45 (2005).

[148] "Self-organization and collective behavior in vertebrates", Couzin, I.D. and Krause, J., **Advanced Studies of Behavior**, 32, 1–75, (2003)

[149] "Chaos–order transition in foraging behavior of ants". Lixiang Li, et al, **Proceedings of the National Academy of Sciences**, May 27, 2014

[150] "Army ants dynamically adjust living bridges in response to a cost–benefit trade-off", Reid, C., et al, **Proceedings of the National Academy of Sciences**, November 23, 2015

[151] "Schooling fishes", Shaw, E., **American Scientist,** 66 (2): 166–175, (1978)

[152] "Fish in larger shoals find food faster", Pitcher, T.; Magurran, A.; Winfield, I, **Behavioral Ecology and Sociobiology**, 10 (2): 149–151 (1982).

[153] "Statistical physics is for the birds", Toni Feder, **Physics Today** 60, 10, 28 (2007)

[154] "The confusion effect when attacking simulated three-dimensional starling flocks", Hogan, **Royal Society Open Science**, 4:160564160564, (January 2017)

[155] "Orientation and Navigation of Migrating Birds", N. S. Chernetsov, **Biology Bulletin**, Vol. 43, No. 8, pp. 788–803. (2016)

[156] "The Cambridge Declaration on Consciousness" Low, P. et al, **Francis Crick Memorial Conference**, University of Cambridge, (7 July 2012).

[157] "Reflections on Dreaming and Consciousness: What is the function of dreaming and how does it operate?" Berezin, R., **Psychology Today**, (October 25, 2013).

[158] "Can machines be conscious?", Koch, C., Tononi, G., **IEEE Spectrum**, vol. 45, no. 6, pp. 55-59, (1 June 2008).doi: 10.1109/MSPEC.2008.4531463.

[159] "Self-consciousness: beyond the looking-glass and what dogs found there", Cazzolla Gatti, Roberto, **Ethology, Ecology & Evolution, 28** (2):232–240. (2015). doi:10.1080/03949370.2015.1102777

160 "Visual discrimination of species in dogs", Autier-Dérian, D., et al, **AnimalCognition,16,** 637–651,(2013). https://doi.org/10.1007/s10071-013-0600-8

161 **Facing the Intelligence Explosion,** Muehlhauser, Luke, e-book, (2013). https://intelligenceexplosion.com/ebook.

162 "The Modern History of Computing", Copeland, B. Jack, **Stanford Encyclopedia of Philosophy**, Metaphysics Research Lab, Stanford University, (18 December 2000). https://plato.stanford.edu/archives/win2020/entries/computing-history/

163 **Science: 100 Scientists Who Changed the World,** Balchin, Jon, **Enchanted Lion Books**, p. 105, ISBN 9781592700172. (2003).

164 **The Hut 6 Story: Breaking the Enigma Codes**, Welchman, Gordon, *Classic Crypto Books*; 2nd edition, (October 19, 1997).

165 "Lecture to the London Mathematical Society on 20 February 1947", Turing, A.M., https://www.vordenker.de/downloads/turing-vorlesung.pdf

166 "Replicating the Manchester Baby: Motives, methods, and messages from the past", Burton, Christopher P., **IEEE Annals of the History of Computing**, **27** (3): 44–60, (2005).

167 **A History of Manchester Computers (2nd ed.),** Lavington, Simon, *The British Computer Society*, ISBN 978-1-902505-01-5 (1998),

168 "Intelligent Machinery", Turing, A.M.,(1948) https://weightagnostic.github.io/papers/turing1948.pdf

169 "Computing Machinery and Intelligence", Turing, A.M., **Mind**, Volume LIX, Issue 236, Pgs. 433–460, (October 1950). https://doi.org/10.1093/mind/LIX.236.433

170 "The Dartmouth College Artificial Intelligence Conference: The Next Fifty years," Moor, J., **AI Magazine**, Vol 27, No., 4, Pp. 87-9, (2006).

171 **The Tumultuous History of the Search for Artificial Intelligence**, Daniel Crevier, **Basic Books**, Simon and Newell, quoted p. 108. (January 1993)

172 Ibid, p. 109

173 Ibid, p. 109

174 Ibid, p. 96

175 Ibid, pp. 158–159

176 **Artificial Intelligence: A Modern Approach** (2nd ed.), Russell, Stuart J.; Norvig, Peter, *Prentice Hall,* Upper Saddle River, New Jersey: (2003). ISBN 0-13-790395-2.

177 **Machines Who Think**, (2nd ed.), McCorduck, Pamela, *A. K. Peters, Ltd.*, (2004). ISBN 978-1-56881-205-2.

178 "AI set to exceed human brain power", Nick Bostrom, **CNN**, (26 July 2006), CNN.com.

179 "IDC FutureScape Outlines the Impact 'Digital Supremacy' Will Have on Enterprise Transformation and the IT Industry**", International Data Corporation** via Business Wire News Releases, (29 October 2019). https://www.idc.com/getdoc.jsp?containerId=prUS45613519

180 "AI Transformation in 2021: In-Depth guide for executives", Cem Dilmegani, **AI Multiple,** (January 1, 2021).

181 "Mathematics Diagnostics/Prescriptive Inventory", Marolda, M.R., **mathdiagnotics.com,** (2020).

182 "Scaling AI in Manufacturing Operations: A Practitioners Perspective", **Capgemini Research Institute**, https://www.capgemini.com/wp-content/uploads/2019/12/AI-in-manufacturing-operations.pdf

183 "A Brief History of Robots in Manufacturing", by Mariane Davids, **RobotIQ**, (July 17, 2017)

184 "Collaborative Robotics: Cobots are Collaborators, AI will make them Partners," Manz, Barry, **Mouser Electronics,** accessed (1 November 2021).

https://www.mouser.com/applications/collaborative-robotics-ai-make-partners/

185 "Swarm Intelligence in Cellular Robotic Systems", Beni G., Wang J., In: **Robots and Biological Systems: Towards a New Bionics?** Dario P., Sandini G., Aebischer P. (eds), NATO ASI Series (Series F: Computer and Systems Sciences), vol 102. Springer, Berlin, Heidelberg, (1993).

[186] "Kilobot: A Low-Cost, Scalable Robot System for Collective Behaviors", Rubenstein, M. et al, **IEEE International Conference on Robotics and Automation**: 3293–3298, (2012).

[187] **Self-Reconfigurable Robots: An Introduction,** Stoy, K, et al., *MIT Press*, (January 2010).   ISBN: 9780262013710

[188] "Automotive Autonomous Vehicles – Global Sector Overview and Forecast to 2035 (Q1 2021 Update)", **GlobalData**, London, UK, Report number GDAUS760321, (March 2021).

[189] "A Detailed Overview of The Trucking Industry", **Transport Rankings**, (2021).

https://www.transportrankings.com/detailed-overview-of-trucking-industry.php,

[190] "U.S. Postal Service Delivers Mail Using TuSimple's Self-Driving Trucks", Alan Ohnsman, **Forbes**, (May,2019)

[191] "Pictures into Numbers", Areste, Romain, **IBM Health Forum**, (September 2, 2020).

[192] Ibid.

[193] "Artificial intelligence and computational pathology", Cui, M., Zhang, D.Y., David Y. Zhang, **Laboratory Investigations**, 101, 412–422, (2021). https://doi.org/10.1038/s41374-020-00514-0

[194] "Detecting cancer metastases on gigapixel pathology images". Liu Y, et al., **arXiv,** Cornell University, (2017).

[195] "The Air Force's AI Brain Just Flew for the First Time," Mizokami, Kyle, **Popular Mechanics**, (May 13, 2021).

[196] "Swarm! Cries the Air Force (And F-35 Pilots Will Love It)", Osborn, Kris, **The National Interest**, (December 11, 2019)

[197] "Boeing's new autonomous fighter jet has a pop-off, swappable nose. Called the "Loyal Wingman," the drone is designed to fly with crewed aircraft." Verger, Rob, **Popular Science**, (15 May 2020).

[198] "Swarm! Cries the Air Force (And F-35 Pilots Will Love It)", Osborn, Kris, **The National Interest**, (December 11, 2019).

[199] "The U.S. Navy Is Building a Swarm Ghost Fleet", Osborn, K., **The National Interest,** (January 24, 2019).

[200] "Swarms of Mass Destruction: The Case for Declaring Armed and Fully Autonomous Drone Swarms as WMD", Kallenborn, Zachary, **Modern War Institute at West Point**, (May, 29, 2020)

[201] "Swarm Control in Unmanned Aerial Vehicles". Hexmoor, Henry, et al, University of Arkansas Computer Science and Computer Engineering Department, http://www2.cs.siu.edu/~hexmoor/CV/PUBLICATIONS/CONFERENCES/IC-AI-05/IC-AI-05.pdf

[202] "KARGU Autonomous Tactical Drone Can Identify and Hunt Human Targets", TechEBlog, (May 31, 2021).

https://www.techeblog.com/kargu-autonomous-tactical-kamikaze-drone

[203] "Swarms of Mass Destruction: The Case for Declaring Armed and Fully Autonomous Drone Swarms as WMD", Kallenborn, Zachary, **Modern War Institute at West Point**, (May, 29, 2020)

[204] "The Global Risks Report -2021, (16th Edition)", Schwab, Klaus, et al, **World Economic Forum**, (2021)

[205] "Some Results on Multicategory Pattern Recognition", Bledsoe, W.W., **Journal of the Association of Computing Machinery**, 13 (2): 304–316, (1966).

[206] "Theory of mind", Ruhl, Charlotte, **Simply Psychology**, (7 August 2020).

[207] "Simulation-Based Internal Models for Safer Robots", Blum, C., et al, **Frontiers in Robotics and AI**, (11 January 2018). https://doi.org/10.3389/frobt.2017.00074

[208] "A technique to estimate emotional valence and arousal by analyzing images of human faces", Fadelli, Ingrid, **Tech Xplore**, (January 26, 2021).

[209] "Operations research's unheralded role in the path-breaking achievement," Chang, Hyeong Soo, et al, **Informs,** Volume 43, Number 5, (2021).

[210] **Metal-oxide Semiconductor Large Scale Integrators Design and Application** (Texas instruments electronics series), Carr, W., Mize, J., *McGraw-Hill Education*, (January 1, 1972).

[211] **Solid State Design,** Vol. 6., *Horizon House,* (1965).

[212] **Nanowire Transistors: Physics of Devices and Materials in One Dimension,** Colinge, Jean-Pierre, et al, *Cambridge University Press. p. 2,* (2016). ISBN 9781107052406.

[213] Ibid.

[214] "Cramming more components onto integrated circuits", Moore, Gordon E., **Electronics Magazine,** (1965-04-19).

[215] "A New Golden Age for Computer Architecture: Domain-Specific Hardware/Software Co-Design, Enhanced Security, Open Instruction Sets, and Agile Chip Development", John L. Hennessy; David A. Patterson, **International Symposium on Computer Architecture,** (June 4, 2018).

[216] **The Age of Spiritual Machines: When Computers Exceed Human Intelligence,** Kurzweil, Ray, *Viking Press,* (1999)

[217] **The Innovator's Way: Essential Practices for Successful Innovation,** By Peter J. Denning, Robert Dunham, *The MIT Press,* (2010)

[218] **Supercomputers: directions in technology and applications,** Hoffman, Allan R.; et al., *National Academies. pp. 35–47, (1990).* ISBN 978-0-309-04088-4.

[219] **Encyclopedia of Parallel Computing,** Padua, David, pgs. 426–429, *Springer,* (2011) ISBN 0-387-09765-1

[220] "The computer as a physical system: A microscopic quantum mechanical Hamiltonian model of computers as represented by Turing machines", Benioff, Paul, **Journal of Statistical Physics,** 22 (5): 563–591, (1980).

[221] "Google and NASA Achieve Quantum Supremacy", Tavares, Frank, https://www.nasa.gov/feature/ames/quantum-supremacy, (23 October 2019).

[222] "On 'Quantum Supremacy'", Edwin Pednault, et al, **IBM Research Blog**, (22 October 2019).

[223] "Quantum supremacy using a programmable superconducting processor", Arute, F. et al, **Nature**, volume 574, pages 505–510, (2019)

[224] "Quantum computational advantage using photons", Han-Sen Zhong, **Science**, Vol. 370, Issue 6523m, pp. 1460-1463, (18 Dec. 2020).

[225] "First Photonic Quantum Computer on the Cloud: Toronto-based Xanadu suggests its system could scale up to millions of qubits", Charles Q. Choi, **IEEE Spectrum**, (09 Sep 2020)

[226] "Xanadu Brings Photonic Quantum Computing To The Cloud", Paul Smith-Goodson, **Forbes**, (Sep 9, 2020)

[227]. "Some Studies in Machine Learning Using the Game of Checkers", A. L. Samuel, **IBM Journal of Research and Development,** 3:3, pp. 210–229, (1959).

[228] "Google achieves AI 'breakthrough' by beating Go champion", **BBC News,** (27 January 2016).

[229] "Learning while searching in constraint-satisfaction problems", Dechter, Rina, **University of California**, Irvine, Computer Science Department, Cognitive Systems Laboratory, (1986).

[230] "Natural Language Processing (NLP)", **IBM Cloud Education,** (2 July 2020). https://www.ibm.com/cloud/learn/natural-language-processing.

[231] Ibid

[232] "Quantum Algorithms for Compositional Natural Language Processing", William Zeng (Rigetti Computing), Bob Coecke (University of Oxford), **Cornell University**, Computer Science, (4 August 2016) arXiv:1608.01406 [cs.CL], https://arxiv.org/abs/1608.01406

[233] "FAA clears Amazon's fleet of Prime Air drones for liftoff", Caine, Aine, **Business Insider**, (31 August 2020) https://www.businessinsider.com/amazon-prime-air-delivery-drones-faa-ruling-2020-8.

[234] "Harvard Is Trying to Build Artificial Intelligence That is as Fast as the Human Brain", Javelosa, J., **Futurism,** (1 January 2016).

[235] "A deep look at a speck of human brain reveals never-before-seen quirks", Sander, Laura, **Science News**, (9 June 2021).

[236] **Superintelligence: Paths, Dangers, Strategies**, Bostrom, Nick, *Oxford University Press*, (2014). ISBN 978-0199678112

[237] "Stephen Hawking AI Warning: Artificial Intelligence Could Destroy Civilization", Osborne, Hannah, **Newsweek**, (Nov 7, 2017).

[238] "The Singularity: A Philosophical Analysis" , Chalmers, David, **Journal of Consciousness Studies,** 17: 7–65, (2010).

[239] **Superintelligence: Paths, Dangers, Strategies**, Bostrom, Nick, Chapter 9, *Oxford University Press.* (2014). ISBN 978-0199678112

[240] **Super-Intelligent Machines**, Hibbard, Bill, *Kluwer Academic/Plenum Publishers*, (2002).

[241] **Superintelligence: Paths, Dangers, Strategies**, Bostrom, Nick, Chapter 9, *Oxford University Press,* (2014). ISBN 978-0199678112.

242"Bostrom on Superintelligence (6): Motivation Selection Methods", Danaher, John, **Philosophical Disquisitions**, (August 18, 2014) https://philosophicaldisquisitions.blogspot.com/2014/08/bostrom-on-superintelligence-6.html

[243] **Superintelligence: Paths, Dangers, Strategies**, Bostrom, Nick, *Oxford University Press*, page 141,(2014). ISBN 978-0199678112

[244] Ibid

[245] "Elon Musk trots out pigs in demo of Neuralink brain implants", Wetsman, N., **The Verge**, (August 28, 2020). https://www.theverge.com/2020/8/28/21406143/elon-musk-neuralink-ai-pigs-demo-brain-computer-interface

[246] "Elon Musk demonstrates brain-computer tech Neuralink in live pigs", Farr, Christina, **CNBC,** (28 August 2020). https://www.cnbc.com/2020/08/28/elon-musk-demonstrates-brain-computer-tech-neuralink-in-live-pigs.html

[247] "Motor neuroprosthesis implanted with neurointerventional surgery improves capacity for activities of daily living tasks in severe paralysis: first in-human experience", Oxley, T.J., et al, **Journal of NeuroInterventional Surgery**, Volume 13, Issue 2, (2021)

[248] "Can Nanobots Kill Cancer Cells?", Thombre, A., **Science ABC**, (9 Jan 2020). https://www.scienceabc.com/humans/can-nanobots-kill-cancer-cells.html.

[249] "DNA storage: research landscape and future prospects", Dong, Yiming, et al, **National Science Review**, Volume 7, Issue 6, June 2020, Pages 1092–1107, (21 January 2020) https://doi.org/10.1093/nsr/nwaa007

[250] "DNA Nanobots Turn Cockroaches Into Living, 8-Bit Computers", Ashley Feinberg, **Gizmodo**, (4/08/14). https://gizmodo.com/dna-nanobots-turn-cockroaches-into-living-8-bit-comput-1560972468

[251] "Scientists pinpoint how many planets in the Milky Way could host life", Rabie, P., **Inverse**, (30 October 2020), https://www.inverse.com/science/how-many-planets-host-life.

[252] "Galactic Habitable Zones", Shige Abe, **Astrobiology at NASA,** (1 November 2021). https://astrobiology.nasa.gov/news/galactic-habitable-zones.

[253] "Discovery boosts theory that life on Earth arose from RNA-DNA mix", Krishnamurthy, R., **Scripps Institute**, (2020). https://www.scripps.edu/news-and-events/press-room/2020/20201223-krishnamurthy-dna.html.

[254] "How an ancient cataclysm may have jump-started life on Earth", Service, Robert, **Science**, (January 10, 2019).

[255] "Searching for Interstellar Communications", Cocconi, G.; Morrison, P., **Nature**, 184(4690): 844–846, (1959).

[256] "The Drake Equation Revisited: Part I, Frank Drake, **Space Dailey,** (29 September 2003). https://www.spacedaily.com/news/life-03zx.html

[257] "W(h)ither the Drake equation?" Burchell, Mark J., **International Journal of Astrobiology**, vol. 5, Issue 3, p.243-250, (December 2006).

[258] "All the Exoplanets That Came Before 1995", Yiu, Y, **Inside Science**, (October 8, 2019).

[259] Ibid

[260] "Exoplanet Exploration", **NASA**, https://exoplanets.nasa.gov.

[261]. "Search for Artificial Stellar Sources of Infra-Red Radiation", Dyson, F., **Science**, 131 (3414): 1667–1668, (1960).

[262] "Alien Megastructure' Ruled Out for Some of Star's Weird Dimming", Wall, M., **Space.com**, (October 04, 2017). https://www.space.com/38363-alien-megastructure-tabbys-star-dust.html.

[263] "How long would it take to colonize the galaxy?", **OpenLearn,** (4 November 2011), https://www.open.edu/openlearn/science-maths-

technology/science/physics-and-astronomy/how-long-would-it-take-colonise-the-galaxy

[264] "The Zoo Hypothesis", Ball, John, **Icarus**, Volume 19, Issue 3, P.347-349, (July 1973).

[265] "The Great Filter – Are We Almost Past It?", Hansen, Robin, (15 September 1998). https://web.archive.org/web/20100507074729/http://hanson.gmu.edu/greatfilter.html

[266] "Aliens Found In Ohio? The 'Wow!' Signal", Krulwich, Robert, **National Public Radio**, (May 29, 2010).

[267] "Dissolving the Fermi Paradox", Sandberg, Anders, et al, **Future of Humanity Institute, Oxford University**, Cornell University, arXiv.org > physics > arXiv:1806.02404, (June 8, 2018)

[268] "Glowing Auras and 'Black Money': The Pentagon's Mysterious U.F.O. Program", Cooper, Helene, et al, **The New York Times**, (Dec. 16, 2017).

[269]"Preliminary Assessment: Unidentified Aerial Phenomena, **Office of the Director of National Intelligence,** (June 25, 2021).

[270] "The Man Who Introduced the World to Flying Saucers", Garber, Megan, **The Atlantic**, (June 15, 2014).

[271] "'Flying Discs' called real by 2 air veterans", Heil, Henry, **Chicago Times,** (7 July 1947).

[272] "RAAF Captures Flying Saucer on Ranch in Roswell Region", **Roswell Daily Record,** (July 8, 1947).

[273] "Pentagon officer and senator say UFOs disabled nuclear capability", Phenix, D., **Mystery Wire,** (May 4, 2021). https://www.mysterywire.com/ufo/uaps-nuclear

[274] "New Mexico 'Disc' Declared Weather Balloon and Kite", **Los Angeles Examiner, Associated Press**, (July 9, 1947).

[275] **Crash at Corona: The Definitive Study of the Roswell Incident**, Friedman, Stanton, Berliner, D., *Paragon House*, (1978)

[276] "UFOs are Real", Friedman, Stanton, writer, Hunt, Edward, Director, **IMDB**, (1979).

[277] "A Roswell requiem", Gildenberg, B.D., **Skeptic**, 10 (1): 60, (2003).

278 "Project Sign and the Estimate of the Situation", Swords, Michael D., **UFO Science**.

https://www.bibliotecapleyades.net/sociopolitica/sign/sign.htm

279 "UFOs, the Military, and the Early Cold War", Swords, Michael, **UFOs, and Abductions: Challenging the Borders of Knowledge,** pp. 82-122, David M. Jacobs, editor; *University Press of Kansas*, (2000).

280 "UFOs, the Military, and the Early Cold War", Swords, Michael, pp. 103 in **UFOs and Abductions: Challenging the Borders of Knowledge,** David M. Jacobs, editor; *University Press of Kansas*, (2000)

281 **The Report on Unidentified Flying Objects**, Ruppelt, Edward, *Doubleday & Co. Inc.*, (1956)

282 "UFOs, the Military, and the Early Cold War", Swords, Michael, pp. 82-121 in **UFOs and Abductions: Challenging the Borders of Knowledge,** David M. Jacobs, editor; **University Press of Kansas**, (2000).

283 "Have we visitors from Space?", Darrach, H.B., Ginna, Robert, **Life** magazine, (April 7, 1952).

284 "When UFOs Buzzed the White House and the Air Force Blamed the Weather", Roos, Dave, **History.com**, (September 6, 2018).

https://www.history.com/news/ufos-washington-white-house-air-force-coverup

285 "50 Years Ago, Unidentified Flying Objects From Way Beyond the Beltway Seized the Capital's Imagination", Carlson, Peter, **The Washington Post**, (21 July 2002).

286 "Memorandum for the Assistant Director for Scientific Intelligence" Durant F.C., **Report of Meetings of the Office of Scientific Intelligence Scientific Advisory Panel on Unidentified Flying Objects**, (January 14–18, 1953, 16th February 1953).

287 **The Report on Unidentified Flying Objects**, Ruppelt, Edward, Doubleday & Co. Inc., (1956)

288 "Report of Meetings of the Office of Scientific Intelligence Scientific Advisory Panel on Unidentified Flying Objects", Durant, F.C., **Memorandum for the Assistant Director for Scientific Intelligence,** (January 14–18, 1953, 16th February 1953).

289 Ibid

[290] **The UFO Book: Encyclopedia of the Extraterrestrial**, Clark, Jerome, pp. 517, Detroit, Michigan: *Visible Ink,* (1998),

ISBN 1-57859-029-9.

[291] Ibid

[292] **The UFO Experience: A Scientific Inquiry**, Hynek, J. Allen, *Henry Regnery Company*, 192-244, (1972).

[293] **UFOs: An Insider's View of the Official Quest for Evidence**, Roy Craig, *University of North Texas Press*, (1995)

[294] **The UFO Book: Encyclopedia of the Extraterrestrial**, Clark, Jerome, *Visible Ink*, pp. 593-604, (1998).

[295] **Final Report of the Scientific Study of Unidentified Flying Objects**, Condon, Edward, Scientific Director, *Bantam Books*, (1968)

[296] Ibid

[297] **UFOs: An Insider's View of the Official Quest for Evidence**, Craig, Roy, *University of North Texas Press*, (1995).

[298] "Preliminary Assessment: Unidentified Aerial Phenomena", Office of the Director of National Intelligence, (25 June 2021).

[299] "FiveObservables", https://fiveobservables.com, (July 4, 2021).

[300] "Ex-CIA director believes UFOs could exist after pal's plane 'paused'", Steinbuch, Yarron, **New York Post**, (April 6, 2021).

[301] "John O. Brennan, Episode 111", Cowan, Tyler, **Conversations with Tyler,** (16 December 2020). https://www.youtube.com/watch?v=LdQ7L0ugZJc

[302] "MT Tech professor claims UFOs are time machines from future" Emeigh, John, KXLF, Butte, Montana, (24 March 2019).

[303] "According to current physical theory, is it possible for a human being to travel through time?", Horowitz, Gary, **Scientific American**, (October 21, 1999).

[304] "Time Travel Is Possible Through Wormholes—but You Can Only Ever Go Backward", Osborne, H., **Newsweek**, (17 November 2017).

[305] **The Many-Worlds Interpretation of Quantum Mechanics**, Everett, Hugh, et al, Princeton Series in Physics. Princeton, NJ: *Princeton University Press.* ISBN 0-691-08131-X.

[306] "Dyson Spheres: How Advanced Alien Civilizations Would Conquer the Galaxy", Tate, K., **space.com,** (January 14, 2014). https://www.space.com/24276-dyson-spheres-how-advanced-alien-civilizations-would-conquer-the-galaxy-infographic.html

[307] "Search for Artificial Stellar Sources of Infrared Radiation". Dyson, Freeman., **Science.** 131 (3414): 1667–1668. (3 June 1960).

[308] "John von Neumann's self-replicating machine — Critical components required," Ellery, A., **IEEE International Conference on Systems, Man, and Cybernetics** (SMC), pp. 000314-000319, (2016).

[309] "It's Easier for Aliens to Visit Us Than Previously Thought", George Dvorsky, **Gizmodo**, (25 July 2013).

[310] "How Self-Replicating Spacecraft Could Take Over the Galaxy", Dvorsky, George, **Gizmodo**, (13November 2013). https://gizmodo.com/how-self-replicating-spacecraft-could-take-over-the-gal-1463732482

[311] "On Oumuamua: Is Oumuamua a Light Sail", Loeb, Abraham, et al, **Harvard University,** (Nov 5, 2018), https://lweb.cfa.harvard.edu/~loeb/Oumuamua.pdf
[312] *"Could Solar Radiation Pressure Explain Oumuamua's Peculiar Acceleration?",* **The Astrophysical Journal Letters**, Loeb, A., et al, (November 12, 2018), https://arxiv.org/pdf/1810.11490.pd
[313] "The Law of Accelerating Returns", Kurzweil, Ray, (March 7, 2001), https://www.kurzweilai.net/the-law-of-accelerating-returns.

[314]. "NASA Releases Plan Outlining Next Steps in the Journey to Mars", Mahoney, Erin, **NASA,** (October 12, 2015).

[315] "Journey to Mars: Pioneering Next Steps in Space Exploration", Mars, Kelli, **NASA**, (August 17, 2016).

[316] "Earth Fact Sheet", Williams, David R., **Lunar & Planetary Science, NASA,** (November 17, 2010).

[317] "Huge Mars Colony Eyed by SpaceX Founder", **Discovery News,** (December 13, 2012).

[318]"SpaceX's Elon Musk teases "dangerous" plan to colonize Mars starting in 2024", Boyle, Alan, **GeekWire,** (June 10, 2016).

[319] "NASA Astronauts Launch from America in Historic Test Flight of SpaceX Crew Dragon", **NASA**, Release 20-057, (May 30, 2020).

[320] . "Goodbye, BFR ... hello, Starship: Elon Musk gives a classic name to his Mars spaceship". Boyle, Alan, **GeekWire**, (November 19, 2018).

[321] "SpaceX's Elon Musk makes the big pitch for his decades-long plan to colonize Mars", Boyle, Alan, **GeekWire,** (September 27, 2016).

[322] "SpaceX's Plans for Mars", Paul Wooster, **21st Annual International Mars Society Convention**, Mars Society, (August 29, 2018).

[323] "Mars colonization: SpaceX thinks Arcadia could be heavenly for a Mars landing", **Canada Journal**, (March 21, 2017).

[324] "Elon Musk puts his case for a multi-planet civilization", Ross Andersen, **Aeon,** (September 30, 2014).

[325] "Death by Magnetic Field: the story of Mars Atmosphere" Xiyu Hugjil Shi, Department of Astronomy, University of Washington, (2016).
https://depts.washington.edu/astron/old_port/TJO/newsletter/spr-2016/xinyu.pdf

[326] "What Is Earth's Average Temperature?", Sharp, Tim, **Space.com,** (April 23, 2018). https://www.space.com/17816-earth-temperature.html.

[327] "Mars Fact Sheet", **NASA,**

nssdc.gsfc.nasa.gov/planetary/factsheet/marsfact.html

[328] "Nasa wants to put a giant magnetic shield around Mars so humans can live there", Beal, Abigail, **Wired**, (3 June 2017).

[329] "Lagrange Points: What are They?" **NASA/WMAP Science Team**, (March 27, 2018).

[330] "Can We Make Mars Earth-Like Through Terraforming?", Mehta, Jatan, **Planetary Society**, (April 19, 2021).

[331] "Inventory of CO_2 available for terraforming Mars", Jakosky, B.M., Edwards, C.S., **Nature Astronomy,** 2, 634–639 (2018).
https://doi.org/10.1038/s41550-018-0529-6

[332] **The Origin of Species**, Charles Darwin, first published 1859, Enhanced Media Publishing, Los Angeles, California, 2017

[333] "What does the fact that we share 95 percent of our genes with the chimpanzee mean? And how was this number derived?", Prescott Deininger, **Scientific American**, (March 1, 2004).

[334] **Lucy's Legacy: The Quest for Human Origins,** Johanson, Donald C.; Wong, Kate, *Harmony Books*, (2009). ISBN 978-0307396396.

[335] "Study Suggests 3.2-Million-Year-Old Lucy Spent a Lot of Time in Trees", Klein, Joanna, **New York Times,** (November 30, 2016).

[336] "Homo habilis", Human Evolution Research, **Smithsonian National Museum of Natural History**, retrieved (31 August 2021) https://humanorigins.si.edu/evidence/human-fossils/species/homo-habilis

[337] "A New Species of the Genus Homo from Olduvai Gorge", Leakey, L.; Tobias, P. V.; Napier, J. R., **Nature,** 202 (4927): 7–9. (1964).

[338] , "Contemporaneity of Australopithecus, Paranthropus, and early Homo erectus in South Africa", Herries, Andy I. R., et al, **Science**. 368 (6486), (3 April 2020).

[339] "Identifying and Describing Pattern and Process in the Evolution of Hominin Use of Fire", Sandgathe, D., **Current Anthropology**. 58: S360–S370, (2017).

[340] "The reputed fossil man of the Neanderthal", King, W. B. R., **Quarterly Journal of Science,** 1: 88–97, (1864).

[341] "Neanderthals Hunted in Groups, One More Strike Against the Dumb Brute Myth", Boissoneault, L., **Smithsonian Magazine**, (June 27, 2018).

[342] "U-Th dating of carbonate crusts reveals Neandertal origin of Iberian cave art.", Hoffmann, D.L., et al, Science, vol. 359, No. 6378, pp. 912-915 (23 Feb 2018).

[343] "Climate Effects on Human Evolution", Human Evolution Research, **Smithsonian National Museum of Natural History**, retrieved (1 September 2021). https://humanorigins.si.edu/research/climate-and-human-evolution/climate-effects-human-evolution

[344] "Homo sapiens", Human Evolution Research, **Smithsonian National Museum of Natural History**, retrieved (31 August 2021).

https://humanorigins.si.edu/evidence/human-fossils/species/homo-sapiens

[345] Ibid

[346] "Population missing and the adaptive divergence of quantitative traits in discrete populations: A theoretical framework for empirical tests", Hendry, A.P., et al, **Evolution**, 55(3), pp. 459–466, (2001).

[347] Ibid

[348] "Physiological and Genetic Adaptations to Diving in Sea Nomads", Ilardo, Melissa, et al, **Cell,** Volume 173, Issue 3, P569-580.E15, (April 19, 2018).

[349] "Can CRISPR–Cas9 Boost Intelligence?", Kozubek, J., **Scientific American**, (September 23, 2016).

https://blogs.scientificamerican.com/guest-blog/can-crispr-cas9-boost-intelligence/

[350] "Brain-Computer Interface, Developed at Brown, Begins New Clinical Trial", Holmer, Mark, **News from Brown**, Brown University, (9 June 2009), https://news.brown.edu/articles/2009/06/braingate2

[351] "Elon Musk's brain-computer interface company Neuralink has money and buzz, but hurdles too", Clifford, Catherine, **CNBC Make It**, (Dec 5, 2020).

[352] " Non-destructive whole brain monitoring using nanorobots: neural electrical data rate requirements", Martins, N.R.B. et al, **International Journal of Machine Consciousness,** Vol. 04, No. 01, pp. 109-140 , *Special Issue on Mind Uploading*, (2012).

https://doi.org/10.1142/S1793843012400069

[353] **The Singularity is Near**, Kurzweil, Raymond, *Viking Press*, (2005).

[354] "Neural-nanorobots to augment human intelligence," Rudin, P., **Singularity 2030**, (26 April 2019),

https://singularity2030.ch/neural-nanorobots-to-augment-human-intelligence

[355] "There's plenty of room at the bottom", Feynman, Richard P., **Engineering and Science magazine**, vol. XXIII, no. 5, (February 1960).

[356] "Nanomaterials and nanoparticles: Sources and toxicity", Buzea, C, et al, **Biointerphases**, 2 (4), (2007).

[357] "Human Brain/Cloud Interface", Martins, Nuno R.B., et al, **Frontiers in Neuroscience**, (29 March 2019)

https://doi.org/10.3389/fnins.2019.00112

[358] "The History of Nanoscience and Nanotechnology: From Chemical–Physical Applications to Nanomedicine", Bayda, S., et al, **Molecules,** 25(1): 112, Published online< (27 December 2019).

https://www.ncbi.nlm.nih.gov/pmc/articles/PMC6982820/

[359] "This Prosthetic Leg is Highly Intelligent", **drive.tech by Maxon,** https://drive.tech/en/stream-content/the-intelligent-prosthetic-leg, (accessed 19 September 2021)